LEARNED ANTIQUITY

Scholarship and Society

GRONINGEN STUDIES IN CULTURAL CHANGE

Volume V

Other volumes in the series:

Volume I: Jan N. Bremmer and Jan R. Veenstra (eds.), *The Transformation of Magic from Late Antiquity to the Early Modern Period*

Volume II: Gerrit J. Reinink and Bernard H. Stolte (eds.), *The Reign of Heraclius (610-641): Crisis and Confrontation*

Volume III: Gillis J. Dorleijn and Herman L.J. Vanstiphout (eds.), *Cultural Repertoires: Structure, Function and Dynamics*

Volume IV: Zweder von Martels and Victor M. Schmidt (eds.), *Antiquity Renewed: Late Classical and Early Modern Themes*

LEARNED ANTIQUITY

Scholarship and Society in the Near-East, the Greco-Roman world, and the Early Medieval West

EDITED BY

Alaisdair A. MacDonald, Michael W. Twomey
and Gerrit J. Reinink

PEETERS
LEUVEN - PARIS - DUDLEY, MA
2003

Illustration on cover: Groningen University Library Ms. Add. 326, fol. 114.
Reproduced with kind permission of the Groningen University Library.

Library of Congress Cataloging-in-Publication Data

Learned antiquity : scholarship and society in the Near East, the Greco-Roman world, and the early medieval West / edited by Alasdair A. MacDonald, Michael W. Twomey, and Gerrit J. Reinink.

p. cm. -- (Groningen studies in cultural change ; v. 5)

"This book is the first of three volumes containing selected proceedings of the international conference on the theme of 'Knowledge, learning, and cultural change,' held at the University of Groningen in November 2001"--Introd.

Includes bibliographical references and index.

ISBN 90-429-1300-2 (alk. paper)

1. Learning and scholarship--History-Congresses. 2. Literature, Ancient--History and criticism--Congresses. 3. Religious thought--To 600--Congresses. 4. Religious thought--Middle Ages, 600-1500--Congresses. 5. Science, Ancient--History-Congresses. I. MacDonald, A. A. (Alasdair A.) II. Twomey, Michael W. III. Reinink, G. J. IV. Series.

AZ301.L43 2003

001.2'09--dc21 2003045507

D. 2003/0602/64

ISBN 90-429-1300-2

CONTENTS

Preface and Acknowledgements vii

Introduction ix

Contributors xiii

The Scientific Revolution of 700 BC
David Brown 1

Berossos and the Mesopotamian Temple as Centre of Knowledge during the Hellenistic Period
Geert de Breucker 13

The Gospels and Cognitive Science
István Czachesz 25

Bridging the Gulf between Transcendence and Immanence in Late Antiquity
Lautaro Roig Lanzillotta 37

Talmudic Tradition in a Changing Society
Wout van Bekkum 53

Between Tertullian and Vincentius Lirinensis: On the Concept *constantia veritatis* and other 'Christian' Influences on Solinus
Zweder von Martels 63

The *Corpus iuris civilis*. Codification and the Encyclopedia of Legal Knowledge
Bernard H. Stolte 81

Between Literary Tradition and Cultural Change: The Poetic and Documentary Production of Dioscorus of Aphrodite
Jean-Luc Fournet 101

Dioscorus and the Law
Peter van Minnen 115

A World Full of the Word: The Biblical Learning of Dioscorus
Jitse H.F. Dijkstra 135

Literary Culture in Sixth-Century Egypt
Teresa Morgan 147

Theology and Medicine in Jundishapur:
Cultural Change in the Nestorian School Tradition
Gerrit J. Reinink 163

Carolingian Bishops and Christian Antiquity: Distance from the Past, Canon-Formation, and Imperial Power
Michael E. Hönicke Moore 175

The Study of Martianus Capella's *De Nuptiis* in the Ninth Century
Mariken Teeuwen 185

Bibliography 195

Index 219

PREFACE

In 1999 the local Research School, Rudolf Agricola, and the Groningen members of the national Netherlands Research School for Medieval Studies succeeded in obtaining a grant for an innovative, large-scale, collective research programme entitled *Cultural Change. Dynamics and Diagnosis*. Supported by the faculties of Arts, Philosophy and Theology and financed by the Board of the University of Groningen, the *Cultural Change* programme constitutes an excellent opportunity to enhance multidisciplinary approaches to phenomena characteristic of transformation processes in the fields of politics, literature and history, philosophy and theology. In order to enhance programmatic cohesion, three crucial 'moments' in European history were selected: 1) Late Antiquity to Early Middle Ages (ca. 200-ca. 600), 2) Late Medieval to Early Modern (ca. 1450-ca. 1650) and the 'Long Nineteenth Century' (1789-ca. 1918). In 2000 and 2002 the two abovementioned Research Schools obtained two more grants for *Cultural Change. Impact and Integration* and *Cultural Change. Perception and Representation* respectively. Several international conferences and workshops have already been organised; more are planned until the end of 2007.

This present volume, number V in the series *Groningen Studies in Cultural Change*, offers a selection of papers presented at the International Conference *Knowledge and Learning* held in November 2001 in Groningen. It is the first of three volumes. The second one (volume VI in the series) entitled *Schooling and Scholarship: The Ordering and Reordering of Knowledge in the Western Middle Ages* will be edited by Alasdair A. MacDonald and Michael W. Twomey. The third one (volume VII in the series) bears the title *Scholarly Environments: Centres of Learning and Institutional Contexts 1600-1960* and will be edited by Alasdair A. MacDonald and Arend H. Huussen. The papers in the present volume, written by scholars from disciplines such as theology, cultural history and law, treat matters of scholarship in Antiquity.

We thank the Board of the University of Groningen for the financial support given to the *Cultural Change* programmes. This initiative has had the happy effect of uniting several faculties and two research schools within a coherent research programme which shows the real advantages of stressing multidisciplinary approaches in the field of the humanities.

The editors are particularly grateful to Marijke Wubbolts for helping to organise the workshop and to Nella Gosman-Scholtens for preparing the texts for publication.

Martin Gosman
General Editor

INTRODUCTION

This book is the first of three volumes containing selected proceedings of the international conference on the theme of *Knowledge, Learning and Cultural Change,* which took place at the University of Groningen from 14 to 17 November 2001. The contributions in this volume deal with a wide variety of topics, ranging from the Ancient Near East to the Early Medieval West. In terms of geography, these contributions take us from places as far away as Iran and Iraq, through the countries surrounding the Mediterranean (Palestine, Syria, Egypt, Asia Minor, Greece, Italy) to the realm of the Carolingians in the West. They bring us into contact with many different cultures and languages: those of Neo-Assyrian, Babylonian and Hellenistic Mesopotamia; of Late Antique and Medieval Judaism; of the Late Antique Greco-Roman world; and of Christianity under the Roman, Byzantine, Islamic and Carolingian Empires.

The multi- and cross-cultural approach adopted here in the study of cultural phenomena and historical developments – one which involves such a wide range of areas and so many periods – is a response to the aims of the large-scale research project on cultural change that is currently running at the University of Groningen. In the context of this programme, therefore, it is appropriate that the central theme of this conference, which in its widest sense concerned potentially all aspects of the history of scholarship, is viewed particularly from the perspective of cultural change. Within this thematic focus, the essays in the present volume discuss not only the contributions of specific centres of learning to cultural change, but also the question of what role was played by particular scholars, literary genres, structures of knowledge and learning, and their socio-political contexts, within the large processes of cultural development and transformation.

The present selected proceedings have been given the collective title of *Learned Antiquity: Scholarship and Society in the Near East, the Greco-Roman World, and the Early Medieval West.* The arrangement of contents is basically chronological, albeit that, within this framework, an attempt has been made to bring together contributions which deal with the same, or closely related, topics, especially where geographical considerations make such juxtaposition desirable. The three contributions dealing with Dioscorus and his work, for example, appear as a cluster. Within this cluster Jean-Luc Fournet's paper deals with some literary aspects of Dioscorus's work. Peter van Minnen's contribution focuses on legal aspects (making a suitable transition from the paper by Bernard Stolte) and Jitse Dijkstra's contribution centres on Dioscorus's scholarship. Equally, the contribution of Gerrit Reinink, concerning the scholarly milieu in the Iranian city of Jundishapur during the ninth century, is for geographical reasons placed after that of Teresa Morgan, which deals with Egypt, and before the papers of Michael Hönicke Moore and Mariken Teeuwen, both of which are concerned with the scholarship at the Carolingian court and with the role of imperial power in the diffusion of knowledge in the Carolingian Empire at the end of the eighth and in the ninth century.

To varying degrees, the contributions reflect the three main subthemes of the conference: Centres of Learning, Encyclopedic Texts and Paradigm Shifts.

Three contributions in particular concern the identification, origin, development and function of centres of learning. In his paper *Berossos and the Mesopotamian Temple as Centre of Knowledge during the Hellenistic Period*, Geert de Breucker discusses the role of one institution, namely the Babylonian temple, as a centre of knowledge in which the traditional Mesopotamian sciences were studied and transmitted up to Hellenistic times. Two cities as centres of scholarship (the Nestorian Christian milieu of Jundishapur and the Islamic milieu of Baghdad) and the cultural interaction between them in the ninth century occupy a central place in Gerrit Reinink's paper *Theology and Medicine in Jundishapur: Cultural Change in the Nestorian School Tradition.* Teresa Morgan's study of the *Literary Culture in Sixth-Century Egypt* concentrates on the cultural conditions and the centres of learning within a selected, larger area, namely Egypt. On the basis of a comparative diachronic and diatopic survey of the Greek and Coptic literary papyri preserved from the first to the seventh centuries, Morgan makes an illuminating estimation of the production of literary culture in sixth-century Egypt.

The formation of enyclopedic texts and their function, authority and tradition under changing historical conditions, are discussed in four contributions. In his paper *The Corpus iuris civilis. Codification and the Encyclopedia of Legal Knowledge*, Bernard Stolte explains the origin and position of the Justinian *Digest* as *the* enyclopedia of legal knowledge, and one which had a lasting role as reference-text in Byzantium and Western Europe, even up to modern times. The practice of legislation in sixth-century Egypt after the promulgation of Justinian's legal corpus is elucidated in Peter van Minnen's contribution on *Dioscorus and the Law.* For its part, Wout van Bekkum's paper *Talmudic Tradition in a Changing Society* takes us to the Jewish communities in the Roman-Byzantine Empire and the Parthian, Persian and Islamic periods (third to seventh centuries). Van Bekkum discusses the role of the encyclopedic text of the Talmud and the authority of the rabbinic sages as a religio-social elite, in relation to the changing political and social conditions and the life of ordinary people of the Jewish communities. The importance of the '*Fundgrube* of ancient knowledge' which circulated in centres of learning of the Carolingian empire – Martianus Capella's work *De Nuptiis*, together with the question of who could have been responsible for equipping the text with its oldest corpus of glosses and for the diffusion of the text together with these glosses in the monasteries in Northern France – are subjects discussed in Mariken Teeuwen's *The Study of Martianus Capella's* De Nuptiis *in the Ninth Century*.

Six contributions deal in one way or another with changes in knowledge and learning, whether these implied either real structural changes, or developments which were induced by 'external' cultural influences, or shifts of interpretation and accentuation within the existing tradition of knowledge and learning. In his paper *The Scientific Revolution of 700 BC* David Brown argues that around that date a genuine paradigm shift took place in the Mesopotamian civilisation of the Neo-Assyrian Empire, namely, the revolutionary change of the ancient science of celestial divination from a science without astronomy to one in which predictive astronomy became a regular constituent. The question of the possible influence of concepts from the Christian *Umwelt* on the Latin grammarian and historian Solinus (second half of the fourth century) is raised by Zweder von Martels in his paper *Between Tertullian and Vincentius Lirinensis: On the concept* constantia veritatis, *and other 'Christian' influences on Solinus*. The influence of the Christian tradition, in particular biblical learning, within a corpus of literary texts is discussed in the third contribution on Dioscorus in this volume. In his paper *A World Full of the Word: The Biblical Learning of Dioscorus*, Jitse Dijkstra notes that among the works preserved in the archive of the Christian notary Dioscorus, biblical quotations and allusions are found almost nowhere except in the latter's petitions to the political authorities. On the basis of an analysis of one specimen, Dijkstra defines and explains the use of biblical learning in this category of documents. The third paper on Dioscorus discusses the topic of this scholar's place in the literary tradition, and it places several related questions in a wider context. Jean-Luc Fournet's *Between Literary Tradition and Cultural Change: The Poetic and Documentary Production of Dioscorus of Aphrodito* focuses on questions of cultural transformation visible within the poems of Dioscorus – such as the alliance and blending of Christian and Classical elements and the interaction between literature and the genre of official document within Dioscorus's literary production. The interpretation, re-interpretation and redefinition of literary tradition and the historical past take a central place in the last two contributions on this subtheme. Lautaro Roig Lanzillotta's article, *Bridging the Gulf Between Transcendence and Immanence in Late Antiquity*, discusses, especially on the basis of Alexander of Aphrodisias's work (c. 200) and the Christian *Acts of Andrew* (second half of the second century), the general tendency, visible in Late Antiquity, to take a mystical approach to Aristotle's writings. In his paper *Carolingian Bishops and Christian Antiquity: Distance from the Past, Canon-Formation, and Imperial Power* Michael Hönicke Moore expounds under which historical circumstances in the Carolingian period a new conceptualisation of the Christian past was able to evolve, and he shows for which purposes imperial and ecclesiastical power accorded a position of eminence to the patristic authors.

István Czachesz's paper, *The Gospel and Cognitive Science*, which discusses the question of how modern insights from cognitive science can be used for the analysis of early Christian (gospel) literature, occupies a special place within the present context. Czachesz's study deals with models of human knowledge in relation to the formation and transmission of oral and written tradition; he shows the relevance of cognitive psychological approaches to knowledge and learning in general, and he brings out the ways in which these new approaches are able to throw a new light on processes of cultural change.

To say that in pre-modern times cultural change occurred throughout the ages and within the different civilisations in the East and West may seem to be stating the obvious. However, defining what cultural change actually is, and discerning where, how and why it took place, is quite another story. Just as culture is a complex, multiform, many-sided and sometimes ambiguous phenomenon, so too is cultural change – and perhaps even more so. The fact that we, more often than not, lack (a great deal of) the historical and contextual information, and in particular the data necessary to make confident diachronic comparisons, is a circumstance that implies several questions: first, what has to be attributed to continuity and what to change; second, where did tradition undergo a real transformation within a confined space of time, and where is it more appropriate to speak of gradual development; and third, which were the social conditions that fostered cultural change or even rendered it imperative? Notwithstanding these restrictions, which are inherent in all research into such early periods, the contributions collected in this volume show that connecting the question of cultural change to the field of knowledge and learning and to some of its relevant literary witnesses is an approach both enriching and rewarding, since it without doubt contributes to the further elucidation of the past, and perhaps even also to a better understanding of the present. The topics treated in this volume, therefore, are not only relevant to specialists in the various fields of study, but are likely to stimulate further investigation of comparable or related themes, by showing which approaches and directions of research may be fruitful in the discussion of the complicated questions connected with cultural phenomena and their mutation, transformation or development in a continually changing world.

CONTRIBUTORS

Wout van Bekkum is professor of Semitic Studies, University of Groningen.

Geert de Breucker is a doctoral candidate in the Department of Semitic Studies, University of Groningen.

David Brown is lecturer in ancient history at University College London and research fellow in Indology at the Free University, Berlin. He is also a member of Wolfson College, Oxford and the Oriental Institute there.

István Czachesz is a doctoral candidate at the Faculty of Theology, University of Groningen.

Jitse H.F. Dijkstra is a doctoral candidate at the Faculty of Theology, University of Groningen.

Jean-Luc Fournet is research fellow at the CNRS, Paris and lecturer in Papyrology, Marc Bloch University of Strasbourg.

Michael Hönicke Moore is Carolyn Grant Fay postdoctoral teaching fellow at Honors College, University of Houston.

Peter van Minnen is assistant professor of Classics and Ancient History, University of Cincinnati.

Teresa Morgan is University lecturer in Ancient History and Fellow and Tutor of Oriel College, Oxford.

Gerrit Reinink is senior lecturer in Aramaic and Syriac, University of Groningen.

Lautaro Roig Lanzilotta is a doctoral candidate at the faculty of Theology, University of Groningen.

Bernard Stolte is professor of Byzantine Law, University of Groningen

Mariken Teeuwen is researcher at the Constantijn Huygens Institute, The Hague; she is also postdoctoral fellow at the Faculty of Arts, University of Utrecht.

Zweder von Martels is lecturer in Latin, University of Groningen.

THE SCIENTIFIC REVOLUTION OF 700 BC

David Brown*

The following offers but a flavour of a period and a discipline that to most people exists as snippets of information remembered from their childhood, or from all-too-brief opening summaries in general books. Hammurapi's Law Code, the Ziggurat of Ur, Babylon's Ištar Gate, the war-like Assyrians – all seem probably strangely other and hopelessly distant. This essay draws in part on my monograph,[1] in which the reader can, hopefully, find all necessary references, but here, at least, I hope to show that the gap between my field and those of others can, to some extent, be bridged.

Assyriology, my field, has itself passed through a number of shifts towards and away from the mainstream of the larger world of scholarship. Early in the century it moved from being a branch of biblical and classical studies on the one hand, and away from the so-called 'pan-Babylonismus' movement, for whom all good things stemmed from Mesopotamia,[2] on the other, to the happier hunting ground of serious isolation, an 'eigene Begrifflichkeit'[3] – a subject unto itself, with its own jargon and methods of approaching that culture between the two rivers. More recently the evidence for and against cultural transmission to the biblical and classical world, and to India, has formed the subject of more measured studies attempting to move beyond the simple juxtaposition of similar motifs to the all-important issues of how and why transmission could have occurred – particularly in the fields of myth,[4] divination,[5] and in the exact sciences.[6] At the same time,

---

* I take the opportunity of thanking the Alexander von Humboldt-Stiftung for funding my current stay in Berlin and my participation in the conference in Groningen, of which this paper formed a part.

[1] Brown, *Mesopotamian Planetary Astronomy-Astrology*.

[2] On this movement, see Larsen, 'The "Babel/Bible" Controversy', and Neugebauer's comments in *A History*, pp. 35-52.

[3] A term coined by Benno Landsberger in his 1926 lecture in Leipzig, reprinted in English in Landsberger, 'The Conceptual Autonomy'. Although the latter remains an important essay, in his 1964 postscript Landsberger, understandably, felt that much that he had said in 1926 was 'one-sided' and 'untenable'.

[4] West, *The East Face of Helikon*, discusses the Greek debt to Mesopotamian myth.

[5] Cryer, *Divination*, treats of the Mesopotamian background to biblical divination.

[6] For the time being see Pingree, 'Legacies', though this opts for an overly positivist approach towards 'borrowings', the revison of which forms the core of my current research. For the transmission of cosmologies, see now Talon, '*Enūma eliš*'. The

a younger generation of scholars has being trying to drag Assyriology into the twenty-first century, approaching the data made available by the previous generations in ways that have long been commonplace in classical, medieval, or biblical studies.[7]

All of this is true for Mesopotamian astronomy, by which I mean no more than the accurate prediction of particular heavenly phenomena. The early pioneers in this field concerned themselves with decipherment, largely ignoring the context in which the famed mathematical-astronomical cuneiform texts were written.[8] They found, with particular parameters and mathematical techniques, that the evidence for transmission to Greece and thence to India in the Hellenistic period was overwhelming, and they left it at that.[9] Studying the evolution of Mesopotamian astronomy *in context* has occupied researchers for the last twenty years,[10] and for the first time evidence from Mesopotamia can make a useful contribution to the very notion of what a science is.[11] No longer is the collection of those mathematical-astronomical texts the concern only of historians of astronomy and mathematics, who felt free to account for those texts on the basis of whatever particular view they held dear as to what makes scientists 'tick'. Now, it is fair to say, an understanding of Mesopotamian astronomy-astrology is contributing to a deeper understanding of Greek and other early astronomical systems,[12] and vice versa. Perhaps, in this context, Mesopotamian achieve-

more general questions of what constitutes the Mesopotamian legacy are the concerns of Dalley, *Legacy*; Bottéro, 'Religion and Reasoning', and the *Melammu* group, based in Helsinki, which deals with the intellectual heritage of Assyria and Babylonia from c. 1300 BC on. For details see now http://aina.org/melammu/intro.html#INTRODUCTION.

[7] E.g., Pongratz-Leisten, *Herrschaftswissen*, deals with god-king communication, especially via divination, within the framework of the 'Tübingen school' of historians and sociologists of religion represented by Burkhard Gladigow, Hubert Cancik and Christoph Auffarth. Veldhuis, *Elementary Education*, addresses the lexical and education material in full cognisance of modern thinking on these matters. See in particular pp. 4-8.

[8] These texts mainly date to the period after c. 350 BC, but many precursors are now attested. For a particular view of the stages through which modern scholarship on ancient cuneiform astronomy has evolved, see Brown, 'Misinformation'.

[9] The evidence for earlier borrowings is more controversial, and is the subject of my current research in Berlin.

[10] I date it to the publication in 1983 of Simo Parpola's commentary to the letters and reports sent by the Assyrian and Babylonian scholars to their royal employers. On these letters and reports, see further below.

[11] This was one of the express intentions of my own monograph. See, in particular, p. 9 and chapt. 5.1.3, entitled 'Philosophical Considerations'.

[12] See, in particular, my forthcoming paper 'Parallel Approaches'.

ments can be permitted to offer a contribution to the general understanding of what culture is, and how it might change.

The astronomical knowledge – the parameters and mathematical methods needed to predict eclipses, say, or the behaviour of the planets – did not appear in Mesopotamia in isolation, but emerged from a background of divinatory practice already ancient by the eighth century BC. It did not come about as a result of the efforts of particular scholars working alone, and who were interested in prediction for its own sake.[13] Quite the contrary, it emerged under the auspices of royal support for a network of experts – diviners who interpreted the meaning to the king and state of particular phenomena, and a bevy of other scholars whose job it was to respond to those interpretations by appeasing through chanting or exorcism those gods who were believed to have sent the signs. We know a great deal from their letters and reports[14] about the education of these scholars, and the pressures of their jobs. We know also that the mathematical methods used to make astronomical predictions evolved over time-spans longer than an individual's lifetime,[15] and we are able to surmise with confidence from the names the scholars gave themselves and their texts, and from what the texts predicted, that the purpose of these mathematical astronomical texts was to predict significant, ominous phenomena.[16]

---

[13] As was so often cited or implied in the works of the historians of astronomy who first deciphered and interpreted these texts in the first sixty years of the twentieth century. It was the position maintained by Neugebauer (*A History*, p. 412), and until his death. He argued that the so-called ephemerides (the astronomical cuneiform table texts) exhibit traits that attest to more than the fulfilling of divinatory needs, and rather to 'intellectual interest'. However, the more intricate, complex texts do not reproduce what happens in reality more accurately. The 'intellectual interest', then, was not in modelling nature better, but in a form of mathematical exegesis, which to my mind has a divinatory purpose behind it. See Brown, *Mesopotamian Planetary Astronomy-Astrology*, pp. 161, 220, 243, and note 381.

[14] Now republished in Parpola, *Letters* and Hunger, *Astrological Reports*.

[15] For example, Steele, 'A Simple Function', the section entitled 'Historical Implications', estimates that approximately one hundred years of eclipse records would be needed to determine a basic model for the variation in the amount by which 223 months exceed 6585 days. This variation is crucial in the prediction of the times of future eclipses, and a more sophisticated function modelling the variation, known today as Ø, formed a central part of the system A lunar mathematical-astronomical texts. Britton, 'Lunar Anomaly', p. 245, estimates that the fully developed System A was in place by c. 350 BC (±25 years), but that work had started on its construction some 75 years earlier, and writes that this makes: 'it unlikely that it was the work of one person', which 'considering its internal consistency ... is a surprise'.

[16] This has been an issue of some debate in recent years. Certainly, the accuracy of the predictions made in the mathematical-astronomical texts (e.g. degrees and seconds of the zodiac) exceeds what is recorded in omen protases (e.g. a planet in a

With this fact in mind, and on the basis of the vast number of divinatory texts and references to divinatory practice in texts recovered from sites in Mesopotamia dating from as early as 2500 BC and to as late as 75 AD, I have no hesitation in characterising divination as a major, perhaps the major, cultural-intellectual achievement of Mesopotamian civilisation.[17] I am content to describe the written exemplars of the discipline as scientific in the first instance, because of the systematic way in which they present data and for the way that ominous scenarios were hypothesised and their significance deduced, while recognising that even this form of divination fails more epistemologically-minded definitions – reproducibility, falsifiability and so forth.[18] The same reservation need *not* be retained for the astronomy that emerged from it, for that was truly a kinematics of singular beauty, with a highly informative content. I see, therefore, in the surviving data, evidence for a paradigm shift from a science of celestial divination without astronomy (accurate prediction), to one with. What I shall attempt to outline now, as the basis of three main topics, is the manner in which astronomical prediction evolved during what was a short period of time around 700 BC.

First, the discovery (if such a word may be used) that particular phenomena can be accurately predicted long in advance is first attested in the works of the scholars working in the employ of the last Assyrian kings, based either at their capital Nineveh, near modern Mosul, or in other major cities of the empire, including those in the south, in Babylonia – the home of divination, particularly celestial divination, and the nation that in 612 BC swept aside Assyrian hegemony of the Near East, and imposed its own upon it. The scholars' predictions are haphazard and inaccurate, but enough remains of their correspondence for us to learn how useful to a diviner was the ability to anticipate those celestial configurations potentially worrying to his employer (the king). We learn, also, how competitive the diviners were with each other, complaining to the king about how ignorant or short-

constellation), but the accuracy was still useful to the diviner, I argue. Also, the requirements of zodiacal divination far exceeded those of omen divination, and (as noted) much of the apparent accuracy of the cuneiform mathematical-astronomical texts was in fact 'perceived precision' caused by a lack of appreciation of error. See pp. 215-219 of my book for a summary of the arguments for the dependence of astronomy on divination, and Rochberg-Halton, 'Between Observation and Theory' and p. 55 of her 'Babylonian Horoscopy', for another view.

[17] For summaries see Bottéro, 'Symptômes, signes, écritures' as well as Cryer, *Divination*, pp. 124-215; Pongratz-Leisten, *Herrschaftswissen*; Brown, 'Astral Divination'.

[18] For a discussion of the scientific nature of celestial divination see my book, chapt. 5.1.3, p. 227ff. See also Cryer, *Divination*, p. 122, note 1.

sighted their colleagues were, and boasting about how eclipses, say, that they had predicted had in fact come about:

> The man who wrote to the king, [my lord] is ignorant. He does not kn[ow] the ... the cycle, (or) the revo[lutions of Venus]. Venus is not y[et] visi[ble]! (Parpola, *Letters*, No. 51:10f).

> An incompetent one can frustrate a judge, an uneducated one can make the mighty worry (Parpola, *Letters*, No. 23:r.3).

> I wrote [to the k]ing my lord [thus]: 'The moon will eclipse, [now] it will not pass by, it will occur' (Hunger, *Astrological Reports*, No. 487:1f).

In Babylon, daily records of planetary and other phenomena were taken, beginning around 750 BC and continuing until the demise of cuneiform itself 800 years later.[19] Although we have reason to suspect that had such records been taken earlier, they would have been destroyed in Antiquity,[20] what is most revealing about those we have is that the oldest ones record their data differently from the later ones, although both are clearly part of the same tradition. Texts older than c. 600 BC (and we have only a few) record less accurately a wider range of data, and in a manner closer to the way in which celestial scenarios are described in the divinatory literature. Without here going into details,[21] the best explanation, in my opinion, of the slight change in recording practice is that between 750 BC and c. 600 BC, at the very latest, those data necessary for astronomical prediction were isolated.

The system of royal patronage of scholars was hardly new in Mesopotamia, so we should not extrapolate solely from the fortuitous discovery of an enormous library and archival collection in Nineveh dating to the seventh century BC. For example, an extensive correspondence between royalty and diviners from the city of Mari on the Euphrates to the south-east of Nineveh survives from a thousand years earlier, and a few references to the

---

[19] Sachs and Hunger, *Astronomical Diaries*, vol III, and Hunger, *Astronomical Diaries*, in which the eclipse and planetary records are edited.

[20] Berossus writes in his *Babyloniaca*, §1.5.1: 'Nabonasaros (King of Babylon c. 745 BC) collected together and destroyed the records of the kings before him in order that the list of Chaldaean kings might begin with him' (after Burstein, *The Babyloniaca of Berossus*, p. 164). I also suggested in my book (p. 205) that some figures located in the oldest eclipse records pertained to values for the excess of 223 months over 6585 days, values which would have taken several generations to determine, taking back the date of the earliest records to before c. 745 BC.

[21] For these, see chapt. 4.2 of my book.

significance of astral phenomena are attested.[22] In both cases, the scholars in Mari and in the Neo-Assyrian capitals drew on what were most likely north-Babylonian traditions of divination, which themselves had borrowed heavily from the literate Sumero-Akkadian culture of the south of what is modern-day Iraq. What was new, however, was the scale of the enterprise under the last Assyrian monarchs, and the particular emphasis they placed on celestial, as opposed to other forms of divination. This, and given what we know of the development in the astronomical Diary and related records from Babylon, and the nature of the predictions made by the scholars at this time (a skill then manifestly in its infancy), and what we know of the scholars' lives, leads me to place the invention of predictive astronomy in the period around c. 700 BC, and to account for it on the basis of the particular, competitive environments fostered by the late Assyrian courts.

Secondly, this point of view leads me to argue against the previously held view (held, for the most part by historians of astronomy and mathematics – it must be noted) that astronomy developed gradually over the millennia. It was previously argued that compositions such as *Mul.Apin*,[23] were examples of 'early astronomy', because they contained a preponderance of particular schemes describing how a celestial phenomenon changed over time – schemes incidentally found in various divinatory writings.[24] These schemes, for example, provided values for the length of daylight, or lunar visibility, in the days of an ideal 360-day year. At first glance they appear astronomical, for they look as if they were intended to model something observable, something in the sky. In fact their astronomical *potential*, from a modern point of view, has been confused with their *intention*,[25] for these schemes woefully misrepresented a reality that was all too apparent to the schemes' creators.

To give one simple example, the schemes assumed that all months last thirty days, something it would not take long to discover was nonsense. Instead, I suggest, these schemes were numerical elaborations of widely held views as to the ideal form of the universe and its behaviour. It was the form the universe was believed to have had when first created, as stated in certain religio-cosmogonical texts.[26] These schemes also drew on the 360-day year,

---

[22] For the publication of the evidence and a discussion, see now Durand, *Archives Royales*.

[23] A two-tablet composition of divinatory schemes and omens, with NA and later exemplars, but of much earlier origin, edited now by Hunger and Pingree (*MUL.APIN*).

[24] For example in tablet 14 of the celestial omen series known by its incipit as *Enūma Anu Ellil*.

[25] Brown, 'Astronomy-Astrology in Mesopotamia', p. 44.

[26] Some are listed in my book, App. 1 §§19-20, the rest discussed on pp. 235-239.

that had been utilised for administrative simplicity from at least as far back as the mid-third millennium BC.[27] They served a very specific divinatory purpose, and that was: they permitted the diviner to compare what he saw with what would have happened if the universe were running according to its ideal form. When reality and the ideal coincided this boded well; if not, it boded ill. This is a new interpretation of early astronomy, and for me it is no surprise that these schemes were imbedded in divinatory and numerologically inspired texts. It also accounts for the continued use of texts such as *Mul.Apin,* the finest exemplar of which in fact dates to the latest period of cuneiform writing, a time when texts that were truly astronomical very accurately modelled all the phenomena described in *Mul.Apin.* At such a time, why would an apparently 'early-astronomical' text have been used, when much more effective astronomical texts were being written?[28] I resist the *ad hoc* explanation that texts such as these were written in the late period merely for antiquarian purposes, and suggest instead that *Mul.Apin* was copied because it continued to have a purpose – its original purpose, a divinatory one.

To summarise, I suggest that astronomy-astrology evolved in Mesopotamia in the following way. Written celestial divination emerged from a background of rural know-how concerning facts such as stars rising heliacally at approximately the same time every year, a rule of thumb that adding one additional month every three years keeps the lunations in approximately the right season, and so forth, in combination with an administratively useful idealisation of the year, and not unrelated notions as to the ideal construction of the universe. The recognition that many celestial phenomena recur, I ascribe to this rural know-how, or 'agricultural star-lore', for what is seemingly never pointed out is that what is periodic is not necessarily predictable, and that knowledge of recurrence does not constitute astronomy, in the sense, which I mean, of accurate prediction. No better example can be provided than that of the month, for here is something that patently recurs, but for which the accurate prediction of whether the next one will last twenty-nine or thirty days is deeply problematic.

The usefulness of this unpredictable periodicity for divination is great, however, for if one makes the *ideal* month thirty days long, then with each lunation a new sign appears, boding either good or ill. We know that the scholars had solved the problem of how accurately to predict the length of

---

Some of this material and other cosmological compositions are discussed in Horowitz, *Mesopotamian Cosmic Geography.*

[27] On this connection see Brown, 'The Cuneiform Conception'.

[28] Sometimes by the same authors, or families of authors. See Britton and Walker, 'A 4th Century Babylonian Model', pp. 110-111.

the month at the latest by 568 BC,[29] and other planetary phenomena somewhat earlier. The advantage to the diviner of being able to pass on such valuable pieces of information to his employer seems to me to be self-evident, and it is alluded to repeatedly in the surviving correspondence between scholars and the last Assyrian monarchs. Knowing in advance when an ominous configuration in the sky would occur meant that the diviner could perform his job of protecting the king better, and also compete better for the king's favour, upon which his livelihood depended. Here then, was the motivation for finding out how to predict the seemingly unpredictable. The great intellectual leap was made. Accurate records were taken at first, and with these, longer periodicities with greater predictive potential were elicited. Astronomy proper – the accurate prediction of certain celestial phenomena – begins, finding its full mathematical expression a couple of centuries later.

The third part of this discussion of the revolution which took place in the years around 700 BC is concerned with the significant cultural changes brought about by the appearance within a short space of time of the new skill of accurate prediction. It could *not* be said, however, that the shift from a divination that had no idea of what signs, what divine messages, were to come, to one that did, had an immediate impact on other divinatory disciplines, or indeed on celestial divination itself. It is true that the zodiac, which was invented no later than c. 470 BC in order to facilitate the prediction of the location of planetary phases, had a certain effect. We know of many texts that combine extispicy and the zodiac, or medical diagnostic divination and the zodiac, for example,[30] but this did not alter the fundamental character of these disciplines as a means of eliciting signs at the moment of observation. In celestial omen divination, at first, much continued as always, for in no way did a predictable event in the sky lose any of its ominous significance[31]. Predictable eclipses, for example, continued to require relevant apotropaic rituals to avert the evil portended (they still do in some parts of the world), though obviously preparations could be made in advance with confidence for the first time.[32]

---

[29] As indicated by the predicted value for *kur* in line 11 of the Diary (see note 19 for publications) which dates to 568 BC. See my book, p. 199, for an explanation as to why this implies that the length of the month was being accurately predicted by this date.

[30] References in Koch-Westenholz, *Mesopotamian Astrology*, p. 164. See also the text published by Heeßel, *Babylonisch-assyrische Diagnostik*, pp. 112-130: 'Der diagnostisch-astrologische Text BM 56605'.

[31] Brown and Linssen, 'BM 134701', p. 155.

[32] *Ibidem*, p. 154.

Even the mathematical-astronomical texts of the latest periods were presented in forms that cohered with those established by the then ancient and revered divination. The following diagram compares the scheme for the ideal length of daylight during the year used in divination, with that used in the mathematical-astronomical texts.

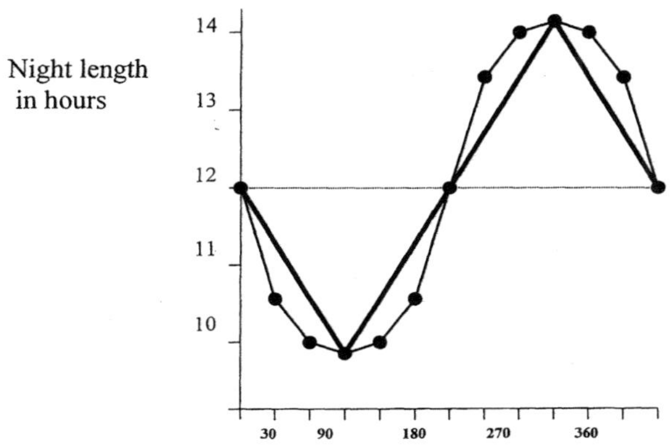

The changing length of the night through the year in:
(a) the mathematical astronomical texts (thin line) expressed as degrees of the zodiac;
(b) the divinatory water clock text BM 29371 (thick line) in days of the ideal year.

In the divinatory text, the ideal value derived from a simple elaboration on the 360-day year was compared, I argue, with what was actually observed in order to derive an interpretation, as strongly suggested by the following two omens, for example:

> If the day reaches its normal length (*ana minātišu ērik*); a reign of long days (Hunger, *Astrological Reports*, No. 7:3).

> If the day is short compared to its normal length; a reign of short days….there will be unreliable speech, bad ways in the land (Hunger, *Astrological Reports*, No. 457:4).

In the mathematical-astronomical text, the changing length of daylight is expressed in terms of the longitude of the sun – in degrees, not days. The ratio of the longest to the shortest day is maintained, but the variation is a closer approximation to a curve. The scheme used in the mathematical-astronomical texts fairly accurately reproduces reality, and yet it preserves the argument of 360 units, split into twelve bands of thirty units, the straightline variation, the evenly distributed solstices, which characterised the divinatory text. It preserves the basic divinatory look, even though its purpose was fundamentally different – namely not to provide an ideal against which observed reality could be judged, but to model reality and make its observation increasingly redundant. This coherence with long-established forms is reminiscent of Ptolemy's use of circular motion when modelling planetary behaviour, as required by the philosophy of Aristotle. As with Ptolemy, the obligation to cohere with forms that were considered basic to the structure of the universe itself did impose certain limitations, although in both cases remarkable ingenuity managed to 'save the appearances', and still produce good astronomy.[33]

Despite the lingering influence of old fashioned divination-of-the-immediate, the new skill of accurate prediction had a profound impact on the *status* of celestial divination, vis-à-vis other forms of divination, without its doing away with the need still to observe significant phenomena. The increasing status of celestial diviners, *ṭupšar Enūma Anu Ellil*, in the seventh century BC reflects this, perhaps, as does the almost total absence of references to the other great diviner – the extispicer, or *bārû*, in texts dating to the Persian and Seleucid periods.[34] The technology of predicting future celestial configurations made celestial divination a particularly attractive discipline for royals, whether local or foreign to Mesopotamia, and the complex and confusing mathematical means by which the predictions were achieved no doubt added an appealing sense of mystery to the whole process, much as the calculations used to write horoscopes today give them an apparent scientific air.

The importance of the skill of astronomy in predicting future scenarios pales into insignificance, however, next to its importance in predicting past ones. From the fifth century BC on, the new technology was used to retro-calculate the state of the heavens at the point of birth or conception, and breathed such life into an old tradition of birth notes[35] that horoscopy was to

---

[33] See my book, pp. 238-239.

[34] McEwan, *Priest and Temple*, p. 15 writes: 'The *bārû* is virtually unattested in texts from the Hellenistic period. This is not surprising since we know that the great vogue enjoyed by astrology at this time made his speciality, divination, superfluous'. This probably overstates the case, since the *bārû*'s job was to check the validity of other interpretations, and this need did not disappear. However, the absence of references to the *bārû* in administrative literature of the late temples is remarkable.

[35] These forms of personal prediction based on events occurring at or near birth are attested in cuneiform sources from the second millennium BC on. For a brief

become the world's dominant divinatory form. Although attested only in some thirty-odd exemplars,[36] horoscopy was of sufficient importance in Mesopotamia itself that the mathematical astronomy developed further to accommodate its particular needs.[37]

The astronomical diaries (see note 19) recorded celestial and other phenomena for the personnel of the temple of Marduk in Babylon on a virtually day-to-day basis, probably from c. 750 BC until at least the middle of the first century BC. We know from the inclusion of unobserved and instead retro-calculated data that it was considered necessary to maintain a *continuous* record. Clearly such a record could have been used to produce a horoscope of any individual born on any date in the past. This may indeed have been part of the purpose of the astronomical diaries, but it is sufficient here to note that their compilers developed the astronomical skills to ensure that missed observations, due to bad weather or whatever, could be replaced by calculated ones. Those skills also meant that the positions of invisible heavenly bodies (those below the horizon at any given moment) could be determined. Before long, it was not the database (which included the record of many unpredictable events, such as meteorological effects) that was necessary for the production of a horoscope, but merely the astronomical and divinatory knowledge. Once this had been acquired, any horoscope could be produced.

It is, perhaps, no surprise that the earliest cuneiform horoscope contained weather data as well as of an astronomical kind,[38] and that all the subsequent horoscopes contained only data that could be retro-calculated. It must be noted, however, that all may have been written up at the time of birth, or at least were based on notes taken at birth. It is a characteristic of all subsequent genathliology of this type, however, that aside from the time and date of birth, little or no observational data were gathered in formulating the first level of interpretation.[39] It was a type of divination, steeped in

---

summary, see Koch-Westenholz, *Mesopotamian Astrology*, pp. 172-173.

[36] See now Rochberg, *Babylonian Horoscopes*.

[37] That was perhaps the purpose of the records and predictions of planet-entry into zodiacal signs in the so-called Goal-Year texts and the Almanacs, and of the so-called 'day-to-day ephemerides', in which the longitude of planets calculated on a daily basis were listed. Normal ephemerides provided only the dates and times of planetary phases. See my book, p. 211 (and note 400).

[38] 'Dense clouds' in line 5 of AO 17649 (text 1 in Rochberg, *Babylonian Horoscopes*).

[39] There is, even today, and unlike what is commonly believed by the sceptic, no one-to-one correspondence between the accurately defined planetary and zodiacal positions in a modern horoscope and their interpretation. The interpretation depends on the 'synthesis' of the hints gained from the astral positions, the expertise of the

history and cloaked in mathematics, which differed from other types in that a large component of the divining process required no observation at all. The tools of the trade were simply knowledge. No animals needed slaughtering, no oil, smoke, birds, noises, astral or atmospheric phenomena needed to be observed, and no records needed to be consulted. All that was needed was a date, a few 'handy tables' of astronomical phenomena, and a working knowledge of the basic premises underpinning celestial divination – the benefic or malefic natures of the planets and the signs, etc. Horoscopy and its related astrological disciplines could and did, therefore, leave the Babylonian temples and spread rapidly via particular experts to Greece and beyond. All subsequent elaborations of horoscopy, including the ascendant or *horoscopus* itself, perpetuated the enviable, non-observed nature of this discipline. Only astronomy, the accurate prediction of certain celestial phenomena, made this possible.

Finally, we may speculate about other, long-term cultural changes wrought by the invention of astronomy in Neo-Assyrian-dominated Mesopotamia – in addition to its underpinning of horoscopes, through the calculation of past phenomena, and its usefulness for eclipse-ritual conference organisers through the calculation of future phenomena. The predictability of phenomena, formerly believed to have been meaningfully altered by gods wishing to inform humanity of their decisions, served to distance the gods from Mesopotamian-based mankind, by establishing the non-arbitrariness of some parts of nature. Is this discovery one of the steps in the gods' gradual withdrawl from all aspects of nature to a position of the first-mover,[40] and the creator of the laws of physics, so far as European thinkers go? Is this not a step in the so-called 'disenchantment' of nature?[41] Was it not this idea that spread from Mesopotamia to the pre-Socratic Greeks – that predictability was possible, for the celestial mechanism could be understood in terms of *how* (by what means) it runs, irrespective of *why* (to what end) it runs?

---

astrologer, and knowledge of the person whose horoscope is being drawn. This knowledge, of course, requires observation.

[40] See my book, p. 234.

[41] Gauchet, *Désenchantement du monde*.

BEROSSOS AND THE MESOPOTAMIAN TEMPLE AS CENTRE OF KNOWLEDGE DURING THE HELLENISTIC PERIOD

Geert De Breucker

Introduction

At the start of the Hellenistic period, around 300 BC, Berossos, a priest[1] of the chief-god of Babylon, wrote a history of his nation and culture. He wrote in Greek, which had become after the conquests of Alexander the Great the *lingua franca* of the Near East.[2] Berossos wanted to address a Greek-speaking readership. He dedicated his work to Antiochus I, the Seleucid king, who ruled over Babylon from 294 until 261 BC.

Although Berossos' work (probably called *Babyloniaca*) is only preserved in fragments, its general contents can be traced. The priest describes the history of Babylonia from the very first beginnings until presumably his own time. It is, however, uncertain whether Berossos also treated astronomical topics in his work.[3] In composing the *Babyloniaca* Berossos used native, cuneiform sources, such as (written versions of) myths, chronicles and king lists.[4]

In this contribution, I treat the following questions: What do we know about the Babylonian temple as a centre of knowledge during the Hellenistic period? Which scholarly professions were connected to it? And, finally, what does that tell one about Berossos?

I start with the first question and present the evidence I have been able to gather about this aspect of the Babylonian temple. The evidence can be

---

[1] The word 'priest' does not imply that Berossos held a cultic or religious function. I would define it as someone who had a connection to the temple. In the Hellenistic period, the Babylonian temple had – apart from being a cultic centre – large economic and political powers; as will be shown, the temple was a centre of knowledge and learning too.

[2] Edition of the testimonies about Berossos and the fragments of his work by Jacoby, *Die Fragmente* (hereafter: FGH). Important studies by Burstein, *The Babyloniaca*; Kuhrt, 'Berossus' *Babyloniaka*'; Verbrugghe and Wickersham, *Berossos and Manetho*, pp. 13-91 (translation and commentary).

[3] The name of Berossos is linked to eight fragments dealing with astronomical/astrological topics (FGH 680 F15-22):

[4] Studies of the cuneiform sources by Drews, 'The Babylonian Chronicles' and Komoroczy, 'Berosos'.

divided in two categories: the archaeological data (the find-spots of tablets) and the texts written on the tablets (their contents). Although in the Hellenistic period the Mesopotamian civilisation was in its decline, and Akkadian and Sumerian were no longer living languages, cuneiform tablets were still written.[5] The tablets represent the traditional genres of the Mesopotamian writing culture: documentary texts (mainly contracts and administrative memoranda), literary compositions (e.g. myths, epics) and 'scientific' texts (e.g. astronomy, astrology, omen texts).[6]

Archaeological evidence

In Assyriology, the disciplines of archaeology and philology were and still today are too often separated. Archaeologists are not really interested in the texts, and philologists disregard the fact that their tablets are archaeological artifacts. The results of both disciplines are not always combined. The merit of the combined approach is clearly demonstrated by Zettler.[7] It is, however, not always possible to study the archaeological context of tablets, simply because this information is no longer available.

This is true for Babylon, Berossos' city, for which we have hardly any archaeological data. Many tablets originating from this city were dug up in illegal excavations. The official excavators of the end of the nineteenth century paid no attention to the archaeological context and specific find-spots of tablets.[8] For the tablets found in more recent campaigns, the archaeological context is known, but these texts are not relevant for our research.[9]

I therefore prefer to turn to Uruk in the south of Mesopotamia. The city was the other main centre of Babylonian cuneiform culture during the Hellenistic period. We have some archaeological information for Uruk at our

---

[5] It should be remarked that it is not always possible to determine whether tablets date from the Achaemenid or Hellenistic period. In these cases, the term 'Late Babylonian' is used.

[6] Oelsner (*Materialien*) has collected the materials up to the beginnings of the 1980s. Since then the corpus of Hellenistic cuneiform tablets has continued to grow.

[7] Zettler, 'Written Documents'.

[8] Oelsner, *Materialien*, pp. 191-194.

[9] The campaigns were undertaken by the Archaeological Project for the Resurrection of the Ancient City of Babylon. Most of the unearthed tablets are school texts dating from the Neo-Babylonian period. In the Temple of Nabû ša Harê, which was then discovered, numerous tablets containing school exercises were found. At least a part of them was dedicated by pupils to Nabû, the god of the art of writing. Therefore, these tablets do not prove that the temple of Nabû was a centre of scribal training; see Cavigneaux, *Textes scolaires*.

disposal. During the regular German excavations of 1959-1960, archaeologists dug in the area of the Bit-Resh, the main temple complex of Hellenistic Uruk, which reached its largest dimensions in that period. They unearthed about 200 tablets and fragments, and 141 copies of cuneiform texts have been published by Mayer and Van Dijk.[10] The corpus of literary[11] compositions consists mainly of texts dealing with the cult (rituals, hymns, prayers) and science (divination, astrology, astronomy and mathematics). The Babylonians considered divination and astrology as science too. They recognised no division between the latter and astronomy; both were branches of the same discipline.

Two historical texts which were found are worthy of mention. The first of these, dating from year 147 of the Seleucid Era (i.e. 165 BC) lists the names of Mesopotamian sages together with the king under whose reign they flourished.[12] The names of the seven antediluvian sages are identical with the names as they occur in Berossos' history.[13] Six of the seven names of the kings also run parallel with the names that Berossos lists. This clearly illustrates the fact that despite the Greek influence, the *Babyloniaca* was a product of cuneiform culture.

The other historical document is a king list.[14] The tablet is broken, but the extant part preserves the names of the Babylonian kings from c. 650 until c. 250 BC (from Kandalanu to Seleucus II). The framework of Berossos' history consisted of king lists.

Our picture of the library of the Bit-Resh is incomplete. Many tablets unearthed in that area were not found *in situ*, but were found scattered round about. Before the official excavations, illegal diggers had already surveyed the area and unearthed texts, which found their way into the modern museums.[15] It is assumed that the texts published by Thureau-Dangin were illegally unearthed in the Bit-Resh, but this remains uncertain.[16] To

---

[10] Mayer and Van Dijk, *Texte*. Only 130 texts were dug up in the area of the temple complex. The others were found elsewhere; see Oelsner, *Materialien*, pp. 143-144.

[11] The term 'literary' is used by Assyriologists to indicate all texts that are not documentary (such as contracts, letters, memoranda). To the category of 'literary' compositions belong e.g. myths and epics, historical compositions (king lists, chronicles etc.), 'scientific' texts (medicine, divination, astrology, etc.).

[12] Van Dijk, 'Die Inschriftenfunde', pp. 44-52; Mayer and Van Dijk, *Texte*, no. 89.

[13] The names can be found in the fragments from Berossos himself and in Abydenus, an obscure historian probably from the second century AD, who used Berossos (edition of the fragments by Jacoby, *Die Fragmente*, pp. 398-410 (FGH 685)).

[14] Van Dijk, 'Die Inschriftenfunde', pp. 53-60; Mayer and Van Dijk, *Texte*, no. 88.

[15] Mayer and Van Dijk, *Texte*, p. 13.

[16] Thureau-Dangin, *Tablettes*. Van Dijk, 'Die Inschriftenfunde' (p. 43) assumes that these tablets originate from the Bit-Resh. This, however, seems unlikely for all the tablets. Two tablets (nos. 17 and 34) contain a colophon with the name of Iqiša.

this group of tablets belong several astrological texts, a medical text and lexical lists. Their contents do not dramatically alter the picture. Neither do the few literary tablets discovered in another important temple of Hellenistic Uruk, the Irigal. The genres represented here are divination, astrology, lexicography, but the number of found tablets is too small to be conclusive.[17]

Since the complete collection of the Bit-Resh (and the Irigal) cannot be recovered, a comparison with tablets found in other Mesopotamian temples is useful. These tablets date from several centuries before the Hellenistic period.

In the area of the Eanna temple at Uruk, the excavators found at least 250 literary texts. They can be dated to the late Neo-Babylonian and early Achaemenid periods (roughly sixth century BC).[18] These tablets were not found *in situ*; they were thrown away as rubbish and possibly even broken on purpose. The material comprises hymns, rituals, hemerologies, epics (e.g. *Enuma Elish*, *Gilgamesh*, a mythological commentary), incantations, medical texts, omens (especially astrological and diagnostical) and lexical lists. Some of the tablets were written by exorcists (*ašipu*).

Literary texts have also been found in the temple of Nabû (god of writing) at Dur-Sharrukin (modern Khorsabad in northern Iraq). This city was built by the Assyrian king Sargon II (721-705 BC) and relinquished after the king's ominous death on the battlefield. The tablets were found in several rooms. Apart from religious texts, omens and lexical lists have been recovered. In one room a tablet with the Assyrian king list was discovered.[19] In two other rooms niches had been made in the wall, and in the niches of one room, a few fragments of clay tablets and inscribed prisms were found. These could be the remains of a library. The niches of the other room were empty.

In the temple of Nabû at Kalhu (modern Nimrud in northern Iraq), another Neo-Assyrian royal city, a tablet storage room has been discovered.[20] Most of the tablets were not found in their original location, but it can be assumed that nearly the complete contents of the library have been recovered. The library was probably in continuous use from 800 BC until the sack of Kalhu (616 BC or later). The original collection consisted possibly

---

Other tablets with the name of this man were not unearthed in the Bit-Resh, but in a residential quarter: see below. Some other, astrological tablets of unknown archaeological context were very probably dug up in the sanctuary; see Oelsner, *Materialien*, p. 144.

[17] Oelsner, *Materialien*, pp. 145, 149-151.

[18] Falkenstein, *Literarische Keilschrifttexte*, pp. 1-27.

[19] Pedersén, *Archives*, pp. 155-158.

[20] On the tablet collection, see Black and Wiseman, *Literary Texts*, pp. 1-7.

of around 300 tablets. Apart from religious texts (prayers, hymns, rituals), a large group of various types of omens is preserved (especially astrological – mostly from the series *Enuma Anu Enlil* – and terrestrial omens), as well as magical and medical texts. Lexical works were also found, as well as a small selection of excerpts from poetic texts (*Epic of Gilgamesh*, *Enuma Elish*, *Lugale*, *Ludlul bel nemeqi*) and proverbs. Elsewhere in the temple, royal inscriptions (e.g. annals) were stored, indicating royal patronage of the sanctuary. Finally, fragments of badly preserved ivory and wooden boards were discovered. A few clay tablets contain a colophon with the name of the scribes. Several copyists were members of a family of exorcists (*ašipu*) and scribes (*ṭupšarru*). A number of these copyists were in royal service, and also fulfilled other duties.

The unearthing of the temple library of the Ebabbar, the temple of Shamash (the Sun God), at Sippar, was spectacular. The library – located in a small room – was discovered during the excavations of 1984-87.[21] In three walls rows of niches had been made, and in the niches the tablets were found, still ranged on their shelves. It is estimated that more than two thousand tablets were stored in the library. The texts date from the Neo-Babylonian and early Achaemenid periods (roughly sixth century BC). Publication of the tablets is still continuing, but we already have a good idea about the contents of the library. It consists of texts dealing with the cult (hymns and prayers), incantations, divination, mathematics, astrology, lexicography and epics (*Atrahasis*, *Enuma Elish*, *Lugale*). A copy of the prologue of the Law Code of Hammurapi, which is a literary composition, was also found, as well as several copies of royal inscriptions and letters. The most ancient king appearing in these texts is Manishtushu (2275-2260 BC). Also mentioned are Hammurapi (1792-1750 BC) and Nebukadnezar I (1126-1105 BC). Most of these texts are fictitious, stating royal privileges to the Ebabbar. Several names of scribes are preserved, among them an exorcist (*ašipu*).

This survey of find-spots of tablets in temples is not exhaustive, but some conclusions can be drawn. It is clear that for the first millennium tablet collections connected with the temple (or temple libraries) can be recovered. To these collections belong texts, which we obviously can ex-pect in a temple, such as cultic compositions (hymns, prayers, etc.). Scien-tific texts – in the Babylonian sense – were also present. The sciences embrace divination, magic, medicine, astrology, astronomy, and mathematics. These sciences had a strong connection with what we would call religious practice. The same is true for epics and myths. The *Enuma Elish* (the 'Epic of Crea-

[21] 'Excavations in Iraq 1984-85', pp. 248-249. Since 1990, al-Rawi and George have been publishing texts from the library in the journal *Iraq*.

tion'), for example, was recited during the New Year's Festival. It is probably useless to make a distinction between texts connected with religion and those which are not when discussing a society in which every aspect of life was influenced by 'religion'. Further, the collection contained historical texts – some of which, such as king lists, had a practical purpose – and lexical lists. These lists were used for training in writing. We know the school curriculum of the late period of Mesopotamian history, but we have no clear idea where the school should be situated.[22] The presence of lexical lists in the archaeological context of the temple is in itself not a sufficient proof that the temple functioned as a school. Some of these texts were dedicated to the temple by pupils who could be taught elsewhere. Other tablets containing lexical lists, however, are not dedicatory and they suggest that the temple – or perhaps more accurately, people connected to the temple – organised school training.[23]

Our reconstruction of temple libraries is, like everything that we can legitimately say about Mesopotamian written culture, based on clay tablets. We know, however, that texts were also written on other materials, such as wooden boards, which did not survive well.[24] Moreover, in the late Babylonian period Aramaic was the spoken language, but most of the texts in this language did not survive, because they were written on perishable material such as parchment. Therefore, our reconstruction will always remain deficient.

Textual evidence

A second category of evidence for the temple as centre of knowledge consists of the texts themselves. Through them we can gain a glimpse of the academic life in practice. Colophons on clay tablets provide interesting evidence of scholarly activities in the temple.[25] Some colophons from the late Babylonian period state that the text was copied from an original belonging to the temple,[26] indicating the existence of a temple library. A number of the

---

[22] Gesche, *Schulunterricht*.

[23] For the distinction between colophons of dedicatory tablets and those of tablets belonging to libraries, see Gesche, *Schulunterricht*, p. 153.

[24] Some archaeological evidence is available (e.g. the remains of boards in the Nabû Temple at Kalhu: see above). Colophons sometimes refer to writing boards as the originals from which the clay tablets are copied; see Hunger, 'Holztafel'.

[25] In Mesopotamia, a colophon could include the following components: bibliographical information (title and subdivision of the copied work; information about the copied text), name of the owner of the tablet, the name of the copyist, a blessing or a curse, a date. Not all components are always present.

[26] For example, Hunger, *Babylonische und assyrische Kolophone*, no. 88 (Uruk:

originals came from other cities,[27] testifying to the intellectual contacts between the various Babylonian cities. These copied texts belong to the same genres as the compositions on the tablets unearthed in temples.

The main scholarly activity displayed in the colophons is the copying, collating, sometimes excerpting, and critical editing of texts. Occasionally, the copyist noted that his original was in a bad state.[28] Some texts are described in the colophon as '*ṣatu u šut pi (ša pi ummani)*' ('commentary and commented words [according to the authority of the scholar]').[29] These texts are of course copies of existing commentaries, but they refer to another activity of the scholars: the studying of and writing of commentaries upon texts. In some colophons people who had a connection with the temple are indicated as owner or copyist of the relevant tablet.[30]

Using the texts, we can discern groups of scholars who had a connection with the temple and its activities. The first profession is that of the *ṭupšar Enuma Anu Enlil* or 'astrologer',[31] for which we have evidence from

---

'original an old wax tablet from the property of (the gods) Anu and Antu'); no. 108 (Uruk: 'property of (the temple) Eanna'); Von Weiher, *Spätbabylonische Texte aus Uruk*, vol. 3, no. 66 (Uruk, 'property of Eanna').

[27] For example, tablets from Uruk copied from exemplars from other cities: Hunger, *Babylonische und assyrische Kolophone*, no. 89 (Borsippa), no. 90 and 106 (Babylon), no. 110 (Nippur); Von Weiher, *Spätbabylonische Texte aus Uruk*, vol. 2, no. 85 (Borsippa); *idem*, vol, 3, nos. 80 and 85 (Babylon).Tablets from Babylon copied from other cities: Hunger, *Babylonische und assyrische Kolophone*, nos. 152, 156 (Bor-sippa).

[28] For example, Hunger, *Spätbabylonische Texte aus Uruk*, no. 94; Von Weiher, *Spätbabylonische Texte aus Uruk*, vol. 2, no. 225, where the scribe noted '*hipi*' (broken) or '*hipi eššu*' (newly broken).

[29] For example, Hunger, *Spätbabylonische Texte aus Uruk*, nos. 31, 33, 49, 50, 51, 72, 84; Von Weiher, *Spätbabylonische Texte aus Uruk*, vol. 2, nos. 38, 54; *idem*, *Spätbabylonische Texte aus Uruk*, vol. 3, nos. 99, 100.

[30] Only a few examples: Anu-aha-ušabši, son of Kidin-Anu, descendent of Ekurzakir was exorcist of Anu and Antu, *šešgallu* (chief sacerdotal officer) of Bit-Resh and astrologer. He owned four tablets of the Epic of *Lugale* (Uruk; Hunger, *Babylonische und assyrische Kolophone*, no. 87); Nidintu-Ani, son of Anu-bel-šunu, descendent of Ekurzakir was exorcist of Anu and Antu. He owned a tablet of the astrological omen series *Enuma Anu Enlil*, which was copied by Anu-aha-ušabši, son of Ina-qibit-Ani, son of Anu-bel-šunu, descendant of Ekurzakir, exorcist of Anu and Antu, *šešgallu* of Bit-Resh and astrologer (Uruk; Hunger, *Babylonische und assyrische Kolophone*, no. 90). Šamaš-eṭir, son of Ina-qibit-Ani, son of Šibqat-Ani, descendant of Ekurzakir was exorcist of Anu and Antu. He owned two astrological tablets, which were written by Anu-aba-uṭer, son of Anu-bel-šunu, son of Nidintu-Ani, descendent of Sin-leqe-unninni, *kalû* (lamentation priest) of Anu and Antu and astrologer (Uruk; Hunger, *Babylonische und assyrische Kolophone*, no. 94).

[31] Literally, the Akkadian name means 'scribe of *Enuma Anu Enlil*'. *Enuma Anu Enlil* is the largest series of celestial and astrological omens. On astrologers and the

Uruk and Babylon.[32] Two texts from the latter city – BOR 4, 132 and CT 49, 144 – describe the tasks of the astrologers.[33] They observe the sky, make the yearly calculations (*terṣetu*) and measurements (*mešhu*) and draw up astronomical texts. *Terṣetu* and *mešhu* refer respectively to mathematical and descriptive astronomical texts.[34] The competencies of the astrologers cover what we should call astronomy as well as astrology; however, as I have already noted, the Babylonians made no sharp distinction between these two disciplines.

The astrologers observed the sky, tried to discover the regularity of celestial phenomena, made calculations and applied their results to human life by making predictions (e.g., horoscopes). The activities of the astrologers resulted in a huge number of tablets relating to various types of astrological and astronomical texts.[35] More than half of the Hellenistic text corpus consists of this kind of texts, and the youngest cuneiform text (75/74 AD) belongs to this category.[36] Babylonians (called Chaldeans by classical authors) were famous in Antiquity for their skills in the field of astronomy/astrology.

An interesting category of texts are the so-called 'astronomical diaries'.[37] In these texts, daily observations of celestial and meteorological phenomena were noted, with, at the end of each month, prices of basic commodities and the water level of the river. Sometimes, the description ends by giving historical information, thus pointing to a connection between astrology/astronomy and 'historiography'. This should make us aware that our categories and concepts about science and disciplines are by no means perfectly adapted to describe Mesopotamian culture.

BOR 4, 132 and CT 49, 144 are protocols based upon a judicial decision taken by the chief administrator and the assembly of Esagila, the main temple of Babylon. In the case of each text, the sons of a deceased astrolo-

---

temple as locus for their activities in the Late Babylonian period, see Rochberg, 'The Cultural Locus' and *eadem*, 'Scribes and Scholars'.

[32] McEwan, *Priest and Temple*, p. 16, lists the names from Uruk (all come from colophons). Boiy, *Laatachaemenidisch en hellenistisch Babylon*, p. 251, gives the names from Babylon. The names occur in documentary texts as well as in colophons.

[33] For an edition, translation and discussion, see Van der Spek, 'The Babylonian Temple', pp. 547-554 and Rochberg, 'Scribes and Scholars', pp. 369-371 and pp. 373-375. The tablets date from the Arsacid period.

[34] Rochberg, 'Scribes and Scholars', pp. 370-371, thinks that 'observing the sky' refers to the genre of the astronomical diaries.

[35] Omens, commentaries on omens, horoscopes, mathematical astronomical texts (mostly ephemerides), descriptive astronomical texts (almanacs, Goal-Year texts, so-called astronomical diaries), handbooks (e.g. MUL.APIN).

[36] Sachs, *Late Babylonian Astronomical and Related Texts*, no. 1201.

[37] Published by Sachs and Hunger, *Astronomical Diaries*.

ger claimed to fulfil the function of their father. After an examination of their abilities in the field of astronomy/astrology, the sons could assume their father's function and collaborate with the other astrologers and their assistants. They received money and arable land from the Esagila. This clearly demonstrates that the astrologers were at the service of the temple.

Another 'scientific' profession which we can link to the temple is that of the *ašipu* (or *mašmaššu*, 'exorcist' or 'incantation priest').[38] This functionary averts and exorcises demons (diseases) by incantations and rituals. Magic had a link with medicine in the sense that both treat sick people.[39] The *ašipu* also participates in purification rituals in the temple. For the Hellenistic period, exorcists are attested in the cities of Babylon, Uruk and Cutha.[40] In Babylon, those who fulfilled the role of *ašipu* received rations from the temple of Esagila.[41] In a fragmentary text an assembly of the exorcists of Esagila is mentioned.[42] Moreover, in Cutha a guild of exorcists is attested, and in one petition this guild requests the chief administrator of the temple of Emeslam in Cutha and the temple assembly to assign arable land to an exorcist, son of another exorcist.[43] Such texts demonstrate the relationship between the function of exorcist and the temple.

By means of the colophons, we can reconstruct (a part of) the private library of one of the exorcists from the Hellenistic period: Iqiša, the son of Ištar-šuma-ereš, descendant of Ekurzakir. Iqiša was an exorcist in Uruk in the early Hellenistic period.[44] The tablets were found in a residential quarter in Uruk. They were, however, not unearthed on their original location, but in a disturbed context. To this day nearly thirty tablets have been recovered. Some colophons state that the tablet was written by other persons – for example, by his son Ištar-šuma-ereš. These scribes were perhaps Iqiša's pupils.[45] Despite this, it is very likely – if not altogether certain – that Iqiša himself wrote the tablets, which were not copied by others.

---

[38] The reading of the Sumerologograms LU$_2$.MAŠ.MAŠ (*ašipu* – preferably – or *mašmaššu*) is debated, see Farber, 'Neues aus Uruk, p. 29 and McEwan, *Priest and Temple*, p. 73.

[39] See the discussion in Finkel, 'On Late Babylonian Medical Training', p. 146.

[40] McEwan, *Priest and Temple*, pp. 21-24. For Babylon, see also Boiy, *Laatachaemenidisch en hellenistisch Babylon*, pp. 249-250. Exorcists occur in colophons, rituals and documentary texts.

[41] CT 49, 126.

[42] CT 49, 140.

[43] BRM 1, 88.

[44] Two colophons have a date: year 6 and 8 of king Philippus Arrhidaeus, respectively 318 and 316 BC (Von Weiher, *Spätbabylonische Texte aus Uruk*, vol. 2, no. 38; Hunger, *Babylonische und assyrische Kolophone*, no. 97).

[45] Hunger, *Babylonische und assyrische Kolophone*, nos. 97, 100, 113, 118; Hunger, *SpätbabylonischeTexte aus Uruk*, nos. 14, 80, 90, 94, 96; Von Weiher, *Spätbabylo-*

The genres represented in the corpus are incantations, omens,[46] astrological/astronomical texts[47] and lexical lists. The latter were certainly used in school training. The similarity between the contents of Iqiša's tablet collection and the aforementioned temple libraries is striking.

Although the tablets of Iqiša were found in a residential quarter and could therefore be defined as a 'private' collection, Iqiša had connections with the temple. In one colophon he is described as 'temple enterer' (*erib biti*) of the gods Anu and Antu.[48] We have other evidence indicating that the boundaries between private and public sphere were fluid in first millennium Mesopotamia.[49] The distinction between private and temple tablet collection is perhaps useless; more important are the people connected to it.

The library of Iqiša reflects the wide range of interests of its owner. Being an exorcist, Iqiša also owned texts relating to other sciences. However, Iqiša was not unique in this respect, and we have evidence of other scholars whose tablet collections displayed a broad range of knowledge.[50] Some scholars combined several functions; for example, that of exorcist (*ašipu*) or lamentation priest (*kalû*) is often combined with that of astrologer (*ṭupšar Enuma Anu Enlil*).[51] Some of these scholars acted as scribes and drew up business contracts between private persons. Such transactions often had a connection with the temple (e.g., sale of prebendary rights or temple property).[52] This too illustrates the imprecise distinction between public,

---

nischeTexte aus Uruk, vol. 2, nos. 6, 9, 18, 21, 25, 32, 34, 35, 37, 38, 43, 44; *idem*, *Spätbabylonische Texte aus Uruk*, vol. 3, nos. 65, 85, 97, 104; idem, *Uruk* (= vol. 4), nos. 147, 158, 159, 188, 220; vol. 5, nos. 245. Two tablets belong to his son Ištar-šuma-ereš: Von Weiher, *Uruk. Spätbabylonische Texte aus Uruk*, vol. 4, no. 141 and vol. 5, no. 240.On the library of Iqiša, see Farber, 'Neues aus Uruk', pp. 26-32.

[46] Various kinds of divination: terrestrial, diagnostic, physiognomic, astrological; text as well as commentaries.

[47] An interesting astrological text is Hunger, *SpätbabylonischeTexte aus Uruk*, no. 94. It contains indications of how to predict prices of grain and related phenomena. The text could be seen as an practical application of the data given in the *Astronomical Diaries*, with their entry about prices of basic commodities.

[48] Hunger, *SpätbabylonischeTexte aus Uruk*, no. 94. On the function of the *erib biti*, see McEwan, *Priest and Temple*, pp. 7-8; Bongenaar, *The Neo-Babylonian Ebabbar Temple*, pp. 147-153.

[49] Oelsner, *Materialien*, p. 145; Bongenaar, 'Private archives', p. 74.

[50] See for example, Pearce and Doty, 'The Activities of Anu-belšunu' (on the tablets owned and/or copied by two lamentation priests from Uruk); Oelsner, 'Aus dem Leben babylonischer "Priester"' (on the tablet collections of *ašipu* and *kalû* in U-ruk).

[51] See the list of McEwan, *Priest and Temple*, p. 16; Parpola, 'Mesopotamian Astrology and Astronomy'.

[52] Anu-bel-šunu, son of Nidintu-Anu, son of Anu-uballit, descendent of Sin-leqi-unninni was astrologer and lamentation priest and his name appears in contracts (as buyer or witness); see Wallenfels, *Seleucid Archival Texts*, pp. 36-37; Pearce and

temple and private sphere. Yet, another 'astrologer' is known to have copied a literary text, the epic of Gilgamesh.[53] The evidence shows that the scholars of the Hellenistic period had – apart from their specific discipline – at least a general knowledge of Babylonian sciences.[54]

Another aspect is that the function of priest/scholar was hereditary.[55] The two protocols, mentioned above, concerning sons who wished to succeed their father as astrologer, illustrate this aspect. It should, however, be remarked that these sons had to pass an examination before they could assume the function. In Uruk, the descendants of Ekurzakir (among whom was Iqiša) held the profession of *ašipu*, whereas the descendants of the Sin-leqi-unninni held the function of *kalû* (lamentation priest).[56]

Berossos and the temple as centre of knowledge

When Berossos states that he is a priest of Bel, the god of the Esagila and patron deity of Babylon, this does not imply necessarily that he had a cultic or religious function; rather it means that he had a connection with the Esagila. Berossos dwelt in a flourishing scholarly milieu, and it is even very probable that he was one of the scholars. As a 'priest', he had at least some knowledge of the traditional Mesopotamian sciences (magic, divination, astrology/ astronomy). This further inclines one to accept the authenticity of the astronomical passages linked with his name.

As a member of the temple community, Berossos worked in the ideal cultural locus in which to write a history of his nation. He had access to age-old archives, or copies of them, and had the expertise to read the traditional cuneiform languages. In his introduction (FGH 680 F1), the priest declares that in Babylon ancient archives were kept with great care. His claim of antiquity is seen to be exaggerated only if we give credence to the impossible year counts; his claim of prudence and carefulness in using his sources is otherwise justified. The activities of the Babylonian scholars in his period and their dealing with tablets and texts prove the veracity of his statement of purpose.

Doty, 'The Activities of Anu-belšunu' (on this Anu-bel-šunu and his grandson bearing the same name). Šamaš-eṭir, son of Ina-qibit-Anu, son of Šibqat-Anu, descendent of Ekurzakir was astrologer, exorcist and *šešgallu* (chief sacerdotal officer). He drew up a sale contract concerning a temple prebend; see Wallenfels, *Seleucid Archival Texts*, p. 38 and McEwan, *Priest and Temple*, p. 16.

[53] Itti-Marduk-balaṭu, son of Iddin-Bel, descendent of Mušezib from Babylon; see Boiy, *Laatachaemenidisch en hellenistisch Babylon*, p. 251.

[54] Oelsner, 'Aus dem Leben babylonischer "Priester"', pp. 235-242.

[55] Kümmel, *Familie, Beruf und Amt*, pp. 156-160.

[56] For a history of the clan of Sin-leqi-unninni, see Beaulieu, 'The Descendants'.

THE GOSPELS AND COGNITIVE SCIENCE

István Czachesz

The purpose of this article is to illustrate how insights from cognitive science can be used for the analysis of early Christian literature; in particular, models of human knowledge will be applied in the study of the problem of the transmission of the gospel tradition. First, certain questions related to the form and genre of gospel literature are mentioned; second, previous attempts at utilising cognitive science for understanding ancient literature are reviewed; third, the use of two cognitive models in the discussion of biblical texts and the gospels in particular is illustrated; finally, I suggest some points for further consideration.

Gospels: title, content, form and genre

Many early Christian writings are titled 'gospels'. In addition to the four canonical gospels (Matthew, Mark, Luke and John), there are a great number of gospels among the apocrypha. To these one has to add the so-called 'Q Gospel', which is beginning to gain the status of a Christian writing of full rights in contemporary research.[1] How the Greek term *euaggelion* (meaning 'good news') became the title of those writings is a complicated issue that we cannot examine in detail now.[2] In some cases, *euaggelion* occurs in the *incipit* of the book, and might have been intended as the title of the text. The gospel of Mark begins with the words 'The beginning of the gospel of Jesus Christ' (Greek: *arche tou euaggeliou Iesou Christou*), and the title of the *Gospel of Truth* comes from the first two words of the text (Coptic: *peuaggelion ntmee*). In both cases, it is probable but not certain that the word 'gospel' refers to the contents of the whole book.[3] In the case of many writings known to us as 'gospels', the title was clearly a later addition to the text.[4]

---

[1] Cf. Crossan, *The Birth*, pp. 31-32, 110-111, 119; Arnal, *Jesus*, p. 1.

[2] Cf. Köster, *Einführung*, pp. 525-526; Guelich, *Mark 1:8-26*, pp. 13-14; Dormeyer, *The New Testament*, pp. 215-219.

[3] For Mark, see Guelich, *Mark 1:8-26*; for the *Gospel of Truth*, see Attridge and McRae, 'The Gospel of Truth', pp. 65-67.

[4] In the first half of the second century AD, *Didache* 15.4 and *2 Clement* 8.5 refer to 'the gospel' probably as a written document, but without attributing it to a particular

Next, we have to ask whether we all the above-mentioned writings can be subsumed under a single 'genre'. Whereas from the four canonical gospels one gains the impression that a gospel is basically a biography of Jesus, other texts that are called gospels show a multiplicity of subjects and forms. Some are indeed biographical, like the infancy gospels,[5] the *Gospel of Peter,*[6] or the *Gospel of Nicodemus.*[7] These texts notably concentrate on either the birth or the death of Jesus. Other gospels, like the *Q Gospel* or the *Gospel of Thomas,*[8] contain only Jesus' sayings. Still other texts, like the *Gospel of Truth*[9] and the *Gospel of Philip,*[10] contain various teachings without directly attributing them to Jesus. Already this brief survey in-dicates that it is problematic to speak of a single 'gospel genre' or even of a separable 'gospel literature' in early Christianity, let alone the evolution of the 'gospel genre'.[11] It is more relevant to ask about the sources and literary models of

---

author. The *Canon Muratori* around 200 AD writes about 'the third Gospel book, that according to Luke' (*tertium evangelii librum secundum Lucas*). Later several writings received the title 'The gospel according to (Greek: *kata*) NN', a usage that was probably not established before the fourth century. Cf. Köster, *Einführung*, pp. 471-472; Schneemelcher, 'Gospels', p. 79; Vorster, 'Gospel Genre'.

[5] The earliest infancy gospel is probably the *Protoevangelium of James* from the second century AD; cf. Culmann, 'Infancy Gospels', p. 419.

[6] Schneemelcher, 'The Gospel of Peter', in: Schneemelcher, 'Gospels', p. 221, dates it to the middle of the second century AD. Recently see Crossan, *The Birth of Christianity*, pp. 481-525.

[7] Also known as the *Acts of Pilate*. The first possible reference to the Acts comes through Justin Martyr (executed in Rome in 165 AD), who twice makes a specific appeal to a writing that he calls the 'Acts of Pontius Pilate' (*Apology* 1.35 and 48).

[8] Valantasis, *The Gospel of Thomas*, p. 13, dates the text between 100 and 110 AD. The dating of the *Gospel of Thomas* largely depends on the hypotheses about its sources. Uro ('Thomas and oral gospel tradition') argues for a the role of secondary orality (oral transmission of written texts) in its composition. In his forthcoming book, Uro further elaborates on the possible interactions of oral and scribal transmission in the *Gospel of Thomas*. I thank Professor Uro for kindly sending me his chapter on 'Orality and Textuality'.

[9] According to Attridge and MacRae, 'The Gospel of Truth', p. 80, the author of the *Gospel of Truth* was familiar with most of the writings of the canonical New Testament. Attridge and MacRae, 'The Gospel of Truth', p. 38, suggest 140-180 AD (provided that Irenaeus, *Against Heresies* 3.11.9 refers to this text).

[10] Dates are proposed from the early second to the third century AD; cf. Schenke, 'The Gospel of Philip'; Turner, *The Gospel According to Philip*, p. 7.

[11] There is no consensus on this matter in contemporary scholarship. Schneemelcher, 'Gospels', pp. 80-81, after surveying the diversity of early Christian gospel literature, yet defends the idea of a 'gospel' genre. Puech and Blatz, 'Other Gnostic Gospels', p. 358, argue that the *Gospel of Truth*, 'despite its title, does not belong to the 'gospel' *Gattung*'. Dormeyer, *The New Testament*, p. 219, tries to solve the problem by talking about 'the genre "kerygmatic ideal biography of Jesus" with the

each text individually. This does not mean that there are not strong relations between some of the gospel texts: the synoptic problem or the *Gospel of Thomas* are notable examples of the opposite. Those connections, however, have to be understood within the larger framework of intertextuality.

In the rest of my article, I want to suggest that the set of problems outlined above can be approached from a new perspective, namely, that of cognitive psychology.[12] I will especially invoke two models of how knowledge is structured in the human mind and highlight some possibilities of how they can help us in analysing the gospel tradition.

Resources of memory in Homer and the gospel tradition

The problem of how long and complex texts like the *Iliad* and the *Odyssey* could be orally transmitted and performed by heart has long ago directed the attention of scholars to the working of memory. Milman Parry and Albert B. Lord studied bards performing epic folk narratives on the Balkan in the first half of the twentieth century, and found that the bards utilise set patterns (phrases and themes) in the narratives.[13] On this analogy, they hypothesised that the Homeric narratives contain 'typical scenes' or 'themes' which were learned by the singers so that they might reproduce an action sequence in song with relative ease: for example, the harnessing of horses, the preparation of a meal, the making of a bed, or the procedures of dressing.[14] This idea was taken up and elaborated on by subsequent Homeric scholarship.[15] Building on the results of the 'oral formulaic school', Elizabeth Minchin has recently turned to cognitive theory for a better understanding of the structure of the Homeric texts. She has analysed larger units of discourse: typical scenes, anecdotes, lists, invocations, descriptive passages, and similes.[16]

---

subgenres "sayings gospel" and "narrative gospel"'. In his classification of the gospels, Crossan, *The Birth*, p. 31 (rightly) proposes 'the vague term, *types*, rather than a more precise term, *genres*', for the classification of gospel literature, 'because there may be several different genres involved in a given type'.

[12] The adjective 'cognitive' is recently applied to a great number of approaches in different disciplinary contexts. This study is inspired by the 'classical' cognitive problems of mental representation and memory, and is limited to some basic concepts which can be readily applied to textual phenomena. For approaches related more closely to the psychology of religion, see Andresen and Forman, 'Methodological Pluralism'; Andresen, *Religion in Mind.*

[13] For 'formulae' and 'themes' in Balkan epic, see Lord, *The Singer*, pp. 30-98.

[14] Parry, 'On Typical Scenes'.

[15] Minchin, *Homer*, p. 6.

[16] Minchin, *Homer*, p. 7.

For the moment let us only briefly summarise Minchin's discussion of two subjects: typical scenes and lists. For the interpretation of typical scenes, Minchin invokes the theory of Robert C. Schank and Robert P. Abelson, who hypothesise that the episodic memory makes use of 'scripts'.[17] Minchin suggests that 'the typical scene (such as dressing, or preparing a meal), which Lord declared to be a chunk of discourse committed to memory by an apprentice singer with performance in mind, may in fact be a scripted entity which encapsulates a standardised record of routine activities and which the singer-to-be, like anyone else, will have learned early in his life. A story is created when typical scenes are linked into a narrative that is always based on a causal chain.[18] The act of storytelling itself follows conventions, which Minchin also describes as a 'script'.[19] Minchin describes in detail the 'contest' script and its variations (tracks) in Homer.[20]

The situation of oral performance, Minchin argues, is crucial for appreciating the curious Homeric lists.[21] Citing contemporary folklore studies, Minchin claims that 'a listening audience will respond to this kind of itemisation with silent attention; and when the list is completed, the performance will often be acknowledged with applause, often louder or more enthusiastic than at any point during the narrative'. Members of the audience are checking the list for its correctness, and they are waiting for the mention of their own family, popular figure, or town.[22] Whereas the recitation of narrative parts gives room for the creativity of the performer, lists must be memorised thoroughly. The text offers mnemonic aids of different kinds: semantic associations of names, geographical arrangement of locations, or auditory aids as rhythm, alliteration, assonance, and the formulaic nature of the epic dialect.[23]

Memory studies have been referred to also in New Testament research. In order to gain a better understanding of early Christian orality, John Dominic Crossan quotes the Balkan studies by Parry and Lord as well as Frederic Bartlett's experiment with the 'the War of the Ghosts'. The bards of Parry and Lord, Crossan argues, transmitted centuries-old folk tradition; Jesus and his first companions, in contrast, dealt with a tradition 'of total newness'.[24] Crossan refers to the famous experiment of Bartlett, in which

---

[17] Both on episodic memory and scripts more will be told below.
[18] Minchin, *Homer*, pp. 14-15.
[19] *Ibidem*, p. 19.
[20] *Ibidem*, pp. 42-72.
[21] For example, *Iliad* 2.494-759, 5.37-83, 18.39-49.
[22] Minchin, *Homer*, pp. 76 and 80.
[23] *Ibidem*, pp. 79-90.
[24] Crossan, *The Birth*, p. 79.

subjects were asked to read a North American Indian folktale and then to reproduce it at divergent intervals of hours, weeks, months, and years. Bartlett found that the reproductions of the tale contained a large number of alterations, which did not only include simplification, rationalisation and omission, but also the introduction of new elements.[25] In conclusion, Crossan warns of the fragility of memory that requires a careful use of the notion of oral tradition.[26]

The studies of Minchin and Crossan are groundbreaking in that they consider the use of cognitive psychology for the study of ancient texts. At the same time, they provoke the student of early Christianity to search the rapidly growing field of cognitive science for further possibilities. For example, Bartlett's above-mentioned experiment lead him to the hypothesis of 'schemas':

> 'Schema' refers to an active organisation of past reactions, or of past experiences, which must always be supposed to be operating in any well-adapted organic response ... Together with the immediately preceding incoming impulse, it [the schema] renders a specific adaptive reaction possible.[27]

That is, we compare our new perceptual input with the schemas in an effort to find something there which is meaningful and familiar. Thus any input which does not fit the existing schemas will either have to be distorted until it does fit, or else in will not be retained at all.[28] Later experiments confirmed that there is considerable difference between the reproduction of something that makes sense for the subject, and something that does not.[29] Consequently, the lesson of Bartlett's experiment is not so much the 'fragility' of memory, as rather the importance of schemas and cultural models for collecting and transmitting information.[30] A particular theory of schemas is provided by Schank and Abelson's scripts, on which also Winchin's study of Homer relies.

Models of memory in biblical literature

In this section, we will experiment with the application of some cognitive models of memory for the understanding of the gospel tradition. According

[25] For the experiment and its evaluation, see Bartlett, *Remembering*, pp. 63-94.

[26] Crossan, *The Birth*, pp. 84 (quoting Daniel L. Schachter) and 85. Cf. Schachter, *Memory Distortion*.

[27] Bartlett, *Remembering*, pp. 201 and 207.

[28] Groome *et al.*, *An Introduction*, pp. 108-111.

[29] E.g., Cole *et al.*, *The Cultural Context*, pp. 267-270.

[30] Cf. Nuckolls, 'Cognitive Anthropology'.

to Endel Tulving, the organisation of knowledge in the human mind relies on two structures: episodic and semantic memory.[31] 'Episodic memory receives and stores information about temporally dated episodes and events, and temporal-spatial relations among these events'. Episodic memories are stored in terms of their autobiographical reference to the already existing contents of the episodic memory store. Let us quote a typical episodic memory: 'Last year, while on my summer vacation, I met a retired sea captain who knew more jokes than any other person I have ever met'.[32] Semantic memory, on the other hand, is 'the memory necessary for the use of language'. 'It is a mental thesaurus, organised knowledge a person possesses about words and other verbal symbols, their meaning and referents, about relations among them, and about rules, formulas, and algorithms for the manipulation of these symbols, concepts, and relations'.[33]

Roger C. Schank and Robert P. Abelson have elaborated on the notion of 'schema' and Tulving's episodic memory, and suggested 'script' as a model for the organisation of memory. 'Some episodes are reminiscent of others. As an economy measure in the storage of episodes, when enough of them are alike they are remembered in terms of a standardised generalised episode which we call a script'.[34] 'A script is a structure that describes appropriate sequences of events on a particular context'.[35] When we receive a sufficient amount of information that is related to a given script, the script is evoked (instantiated).[36] According to Schank and Abelson, 'understanding is knowledge based ... The actions of others make sense only insofar as they are part of a stored pattern of actions that have been previously experienced'.[37] The interpretative power of scripts lies in that once a script is evoked, it supplies information that is not immediately available in the actual situation.[38]

If we try to apply the distinction of episodic and semantic structures to biblical literature we can find candidates for both patterns. For example, in Jewish Scriptures we can identify an 'exodus script', relating how Yahweh released the people of Israel from Egypt. This script is implemented by a number of passages, of which we quote the historical summary of Deuteronomy 26.5-9:

---

[31] Tulving, 'Episodic and semantic memory'; *idem*, 'Memory'; Tulving and Craik, *The Oxford Handbook of Memory*.

[32] Tulving, 'Episodic and semantic memory', pp. 385-386.

[33] *Ibidem*, p. 386.

[34] Schank and Abelson, *Scripts, Plans, Goals and Understanding*, pp. 10 and 16-19.

[35] *Ibidem*, p. 41.

[36] *Ibidem*, pp. 46-50, describe very precise rules for the application of scripts.

[37] *Ibidem*, p. 67.

[38] *Ibidem*, p. 41.

> A wandering Aramean was my ancestor; he went down into Egypt and lived there as an alien, few in number, and there he became a great nation, mighty and populous. When the Egyptians treated us harshly and afflicted us, by imposing hard labor on us, we cried to the LORD, the God of our ancestors; the LORD heard our voice and saw our affliction, our toil, and our oppression. The LORD brought us out of Egypt with a mighty hand and an outstretched arm, with a terrifying display of power, and with signs and wonders; and he brought us into this place and gave us this land, a land flowing with milk and honey. So now I bring the first of the fruit of the ground that you, O LORD, have given me.[39]

On the analogy of the exodus script, we can speak of a 'gospel script' in early Christian literature, the earliest witness of which is 1 Cor. 15.1-8:

> Now I would remind you, brothers and sisters, of the good news that I proclaimed to you, which you in turn received, in which also you stand, through which also you are being saved, if you hold firmly to the message that I proclaimed to you – unless you have come to believe in vain. For I handed on to you as of first importance what I in turn had received: that Christ died for our sins in accordance with the scriptures, and that he was buried, and that he was raised on the third day in accordance with the scriptures, and that he appeared to Cephas, then to the twelve. Then he appeared to more than five hundred brothers and sisters at one time, most of whom are still alive, though some have died. Then he appeared to James, then to all the apostles. Last of all, as to one untimely born, he appeared also to me.

The gospel script is implemented in a number of passages of early Christian literature, the best-known examples being the passion and resurrection narratives. It is possible to talk about a 'martyrdom script', which is implemented in the narratives about the trial and death of several early Christian heroes, beginning with Jesus. Other scripts that can be easily identified include miraculous healing and divine call. The implementation of scripts is triggered by the occurrence of typical motifs. For example, the mention of the execution of a martyr invokes the whole script of arrest, trial, bold confession of faith, and execution, and the possible occurrence of other details. Anyone acquainted with that script will not think, for example, that the hero was assassinated.

---

[39] Biblical passages are quoted after *The Holy Bible Containing the Old and New Testaments. New Revised Standard Version.*

Our application of 'scripts' immediately raises an important question: Where do these scripts come from? Are they generalisations based on life experience like the scripts described by Schank and Abelson? Did anyone ever have a repeated experience of the exodus from Egypt that resulted in the fixation of the script that is widely quoted in Jewish literature? Or did anyone have a repeated experience of the events summarised in the 'gospel' script? The same question can be posed, I think, in connection with the Homeric texts: did anyone have a repeated experience of the gods' taking counsel and then returning to the battle field to direct the course of the war? This is, however, as typical a scene as the dressing of the hero, or the chariot race, a Homeric script that Minchin analysed in detail.

Moreover, many of us seem to possess these scripts as well, since we make good sense of those narratives. Of course, we will have acquired the necessary scripts at school, at Sunday school, from movies, or through reading. It seems that many of our scripts do not originate in 'everyday life' but rather in other forms of cultural exchange. It is very plausible that the same was true for the performers and listeners of Homeric texts as well as for the writers, readers, performers and listeners of Jewish and early Christian texts.[40] This brings us to the central issues of early Christian tradition and Crossan's caveats about the fragility of memory. It must have taken a while until a 'gospel script' or a 'martyrdom script' began to take shape among Christians. And as soon as a given text crossed social or cultural borders (either in an oral or a written form) the scripts of the new environment were different from the ones that shaped the earlier form of the tradition. Since the differences were not as sharp as between Bartlett's 'Red' Indians (from whom the 'War of the Ghosts' originated) and his Cambridge students (who were asked to recall it), the case of complete miscommunication certainly did not take place. I assume that analogies from cognitive anthropology with regard to the functioning of scripts may be of assistance for gaining a clearer picture of the mental circumstances that enabled and determined textual transmission among different groups of early Christians.

Now let us consider examples of semantic memory. Semantic structures of knowledge are easily detectable in wisdom literature: 'Three things are too wonderful for me; four I do not understand: the way of an eagle in the sky, the way of a snake on a rock, the way of a ship on the high seas, and the way of a man with a girl'.[41] The sayings-material attributed to Jesus often assumes similar forms: 'Jesus said, "Where there are three gods, they

---

[40] Parry, 'On Typical Scenes', pp. 405-406, points out tradition as the source of the Homeric 'typical scenes'.
[41] Prov. 30.18-19.

are gods. Where there are two or one, I am with him"'.[42] Another example: 'Then Peter came and said to him, "Lord, if another member of the church sins against me, how often should I forgive? As many as seven times?" Jesus said to him, "Not seven times, but, I tell you, seventy-seven times".[43] These passages represent different literary forms (wisdom sayings and *chreiai*) used in different literary genres (sayings collections and narrative gospels), but the play with numbers provides the cognitive structure of all of them. The pattern occurs in the Jewish *m^{e}shalim*, which category itself designates a cognitive pattern rather than a genre.[44] For the use of numbers in *m^{e}shalim*, in addition to the wisdom-saying quoted above from Proverbs, consider this riddle in 1 Sam. 18.7: 'Saul has killed his thousands / and David his ten thousands'.

Genealogies (Hebrew *toledot*) constitute another form of hierarchical memory. They are frequent in the Old Testament,[45] and also occur in the gospels of Matthew and Luke.[46] From the cognitive point of view, genealogies are lists, similar to the Homeric ones. Other lists in Biblical texts include the blessings of Moses and Jacob, surveys of the lands allotted to the twelve tribes, the description of the making of the Tabernacle, or Ezekiel's vision of the temple.[47] Notice that the 'gospel script' as Paul quotes it also contains a list in which numbers too play an important role:

> ... he appeared to Cephas, then to the twelve. Then he appeared to more than five hundred brothers and sisters at one time, most of whom are still alive, though some have died. Then he appeared to James, then to all the apostles. Last of all, as to one untimely born, he appeared also to me.

Lists are structured with the help of the same methods that biblical scholars have identified in the construction and transmission of sayings-materials: catchwords, thematic and formal groupings.[48] We can add the use of numbers (see above) and initials,[49] as well as the methods identified by Minchin in Homeric lists: semantic associations of names, geographical arrangement

---

[42] *The Gospel of Thomas* 30 (transl. Lambdin, p. 130).

[43] Matt. 18.21-22.

[44] A *mashal* can be a riddle, an insult, a taunt, a curse, a parable, an epigram, or a humorous saying; cf. Ben-Amos, 'Folklore'.

[45] Wilson, *Genealogy*.

[46] Matt. 1.1-17; Luke 3.23-38.

[47] Gen. 33 and 49; Ex. 36.8-39.31; Josh. 13-21; Ezek. 40.48-44.3.

[48] For example, Kloppenborg, *The Formation of Q*, pp. 310-311, 323.

[49] Numbers and initials were typical mnemonic aids in rabbinical tradition; cf. Bowker, *The Targums and Rabbinic Literature*, p. 40ff.

of locations, or auditory aids as rhythm, alliteration, and assonance (see above).

Lists seem to have played an important role in Hellenistic education. Recently Teresa Morgan has called attention to the distinguished position of lists in the Greek school papyri found in Egypt:

> Wordlists survive in both teachers' and pupils' hands. They include all sorts of words: the months of the year, rivers, trades, animals, proper names, even non-sense words, or a mixture of all. Names are evidently so important that they appear from the earliest stages of education. Further, all sorts of proper names – well known and obscure, mythical, literary and historical – are characteristically mixed up together in these texts.[50]

It is even more interesting that this was by no means an introductory exercise which was abandoned when the pupils began to read Homer: 'A large number of texts [copied from Homer] contain parts of lists. In these texts individual lines are typically taken from a continuous passage to form, effectively, a list of names or terms; occasionally a passage which contains a large number of names or terms is copied complete'. Among others, there is an excerpted version of the enumeration of the ships in Book Two, lists with the parts of a boat, with the elements of armour, with the names of gods, nymphs, and heroes. Other texts collect Homeric similes and metaphors, portents, and invocations to the gods. Battle scenes are also quoted in list-like passages in which one hero after another is brought to earth.[51]

Following an overview of the literary texts in the school papyri, Morgan concludes: 'Whatever the pupil was learning, he was evidently not acquiring the bank of shared stories, myths and traditions which for centuries had helped to define Greek identity'.[52] In general, the papyri 'summarise rather than display Greek culture, favouring lists of names above stories, the beginnings of texts and a few well-known phrases above passages distinguished for their coherence or literary qualities'.[53] Morgan warns that literate education in provincial Egypt was different from its counterpart among elite *literati* at Rome or Alexandria or Athens. However, for the study of early Christianity, we have to understand precisely the cognitive schemas in non-elite (and provincial) Hellenistic cultural environments.

---

[50] Morgan, *Literate Education*, p. 101.
[51] *Ibidem*, pp. 107-108.
[52] *Ibidem*, p. 108.
[53] *Ibidem*, p. 118.

Another group of school papyri contains gnomic sayings.[54] The area of gnomic literature has already attracted the attention of New Testament scholarship.[55] For the moment, let us only notice that these texts also witness to the prevalence of semantic memory over narrative scripts. The examples quoted by Morgan suggest that the use of wisdom collections in early Christianity has to be understood in the larger framework of the dominance of lists (and semantic approaches) in Hellenistic education. palaeographical evidence from Roman Palestine suggests that Jewish education was based on similar methods.[56]

Narrative and semantic knowledge structures often appear embedded in each other. For example, whereas a healing performed by Jesus relies on the 'healing script' as far as its narrative organisation is concerned, it also evokes semantic knowledge about God, demons, illnesses, healing powers and techniques, geographical locations, and customs. Paul's summary of the gospel that we quoted above contains a list of the persons to whom Jesus appeared. Both scripts and semantic knowledge structures refer to a number of other scripts and frames. The deeper one understands a given text the more scripts and frames in one's mind are activated.

Some issues for further consideration

In this study I have been able to mention only a few aspects of cognitive science that may be fruitfully used in the research of early Christian literature and the gospel tradition in particular. In the final part of this article I wish to call attention to some possible tasks of future research.

a) The models of cognitive science can be applied to the analysis of the overall design of texts as well. Instead of setting up taxonomies of the gospels, one can examine how different cognitive models interact within them. It is possible that some texts do not follow an overall compositional strategy, but are rather organised into several narrative and semantic substructures.

b) Recently Martha Lee Turner has argued that the structure of the *Gospel of Philip* should be understood in the context of ancient private notes.[57] She based her theory partly on the early Christian use of codex, an object that was known first of all as the material of ephemeral writings.[58] The same perspective has been raised in connection with the *Gospel of*

---

[54] *Ibidem*, pp. 120-151.

[55] For example, Kloppenborg, *The Formation of Q*, pp. 263-316.

[56] Hezser, *Jewish Literacy*, pp. 87, 331-340.

[57] Turner, *The Gospel According to Philip*, pp. 60-79.

[58] Gamble, *Books and Readers*, pp. 49-50.

Thomas.[59] The study of early Christian literacy has to be connected to the study of Jewish (and) Hellenistic sub-elite literacy, examples of which have been mentioned in this study.[60] This research may also be situated in the framework of cognitive psychological approaches to knowledge and learning, which might further improve our understanding of oral and written communication among early Christians.

c) Cognitive studies also provide invaluable insights about bilingualism.[61] This situation was typical of most early Christian groups, which used on different levels Greek and Aramaic (also Hebrew), Greek and Syriac, Greek and Coptic, or Greek and Latin.[62] It is an especially important task to understand education in such an environment.[63]

d) Finally, the examination of knowledge structures should by no means be restricted to the realms of orality and literacy. In his classical study, Lévi-Strauss described how the settlement structure, hair-style, clothing and dances of Bororo Indians contain knowledge about themselves and their universe.[64] Consequently, a cognitive approach to early Christian tradition has to take into consideration factors such as social system and religious rites, down to the level of housing, clothing, eating, and body-language.

Whereas many of these problems have been discussed in New Testament scholarship, approaching them from the perspective of cognitive science may considerably enhance our understanding of the gospel literature as well as of early Christianity in general.

[59] Uro, 'Orality and Textuality' (see note 8 above).

[60] For reading as an upper-class activity in the first and second centuries AD, see recently Johnson, 'Toward a Sociology of Reading', pp. 612-624.

[61] For a survey of recent cognitive approaches to bilingualism, see Wei, *The Bilingualism Reader*.

[62] Cf. Horsley, *New Documents*, pp. 5-40.

[63] Cf. Hezser, *Jewish Literacy*, pp. 90-94, 103.

[64] Lévi-Strauss, *Anthropologie structurale*, pp. 133-180, 268-299; *idem*, *Tristes tropiques*.

BRIDGING THE GULF BETWEEN TRANSCENDENCE AND IMMANENCE IN LATE ANTIQUITY

Lautaro Roig Lanzillotta

διὸ καὶ ἀρχὴ καὶ τέλος νοῦς
Arist., *EN* 1143b

According to a surprising statement by E. Zeller in his monumental *Die Philosophie der Griechen*, the interpretation of Alexander of Aphrodisias would be responsible for the elimination of the 'mystical element' from Aristotle's philosophical system.[1] A century later, however, the opposite opinion would be held with equal conviction by G. Reale: Alexander would be responsible for the introduction into Aristotelianism of alien mystical tendencies, falsifying in this way the scientific and empiricist spirit of Aristotelian philosophy.[2] It is precisely the alleged falsification, of which both opinions accuse Alexander, that would explain the exhaustion and final disappearance of Aristotelianism in late Antiquity. Since no relevant new textual discoveries have been made in the last century to explain such contradictory statements, the alleged forgery seems rather to be a scape-goat for the frustrated expectations of interpreters in reconstructing their own preconceived Aristotle.

The purpose of the present contribution, however, is not to claim an empiricist or mystic construction of Aristotle.[3] Rather I shall explore the general tendency in Late Antiquity (Alexander included) to approach Aristotle's writings mystically. With this purpose I review certain Aristotelian issues that recurrently appear in writers of Late Antiquity. This survey shows that whether or not this was the original intention of those issues, they appear wholly integrated in the philosophical thought of the period in order to serve its marked religious tendencies. I firstly focus on Alexander's interpretation of Aristotle's noetics and his development of the so-called 'rationalistic mysticism'. This mysticism of reason, as defined by Philip

---

[1] Zeller, *Die Philosophie*, p. 793. Cf. Merlan, *Monopsychism*, p. 34.
[2] Reale, *A History*, p. 33.
[3] For the issue, see Merlan, *Monopsychism*, pp. 30-33 and note 2. For the influence of the Jaegerian evolutionist theory on the interpretation of Aristotelian thought and for new insights on the issue, see Bos, *Cosmic and Meta-Cosmic Theology*, pp. 97-112.

Merlan, describes, on the one hand, the 'theory teaching that the highest moments of man's existence are those of his absorption into whatever he takes the divine to be'. On the other, it 'designates a condition of man's mind which could be called supremely rational rather than irrational'.[4] Secondly, a brief survey of other writers of the period exemplifies the recurrent appearance of Aristotelian elements especially in those contexts that state man's transcendent origin and claim that the reunion with the divinity is the main goal of his life. Finally, I consider possible Peripatetic echoes in the *Acts of Andrew*, an early Christian text that conceives the return to God as a deconstruction of the cognitive process. This deconstruction or reversal has the intention of regaining an original intuitive apprehension concomitant with the reunion of subject and object in the act of knowing.

Alexander

Alexander's differentiation between three sorts of νοῖ or 'intellects', namely the 'material intellect' (ὑλικὸς νοῦς),[5] the 'acquired intellect' (ἐπίκτητος) or intellect *in habitu* (ἐν ἕξει),[6] and the 'productive intellect' or νοῦς ποιητικός[7] basically relies on an interpretation of *de Anima* III.5. Aristotle indeed distinguishes in this *passus* the passive intellect (νοῦς παθητικός), which becomes all things by being acted upon, from the active intellect that makes all things (ὁ νοῦς τῷ πάντα ποιεῖν).[8] However, whereas Aristotle restricts himself to applying the notions of potentiality and actuality to the human soul,[9] Alexander applies them to the νοῦς and introduces a divine productive intellect (νοῦς ποιητικός),[10] which coming into man from without is the cause (αἴτιος) of the other two turning into actuality. Nevertheless, in doing so Alexander relies exclusively on Aristotle. Firstly, he divides Aristotle's passive intellect into two, namely the material intellect and the acquired intellect, on the basis of the opposition of potentiality and actuality. Whereas the former might be compared with the 'unwrittenness' of a blackboard because of its receptiveness,[11] the latter,

---

[4] Merlan, *Monopsychism*, pp. 1-2; 20-21. Since Merlan's interpretative breakthrough, his position has originated certain opposition. For a state of the subject see Sharpless, 'Alexander of Aphrodisias', pp. 1224-1225.

[5] Alex. Aphr., *de Anima* 81.9-28.

[6] Alex. Aphr., *de Anima* 82.1-6.

[7] Alex. Aphr., *de Anima* 88.23-89.12.

[8] Aristotle, *de An.* 430a 10-19.

[9] Aristotle, *de An.* 430a 13-14; Cf. Donini, *Tre studi*, p. 38ff.

[10] See *infra*, note 45.

[11] Alex. Aphr., *de Anima* 84.24-26: 'Unwrittenness' (*tabulae rasum*), transl. Merlan (*Monopsychism*, p. 14).

once it has been instructed, is the form and perfection of the former.[12] Secondly, Alexander draws on Aristotle's differentiation between two sorts of actuality in *de Anima* II.1, which in the case of knowledge can be compared with the possession of knowledge and the active exercise thereof.[13] Only the active exercise can be considered as intellect κατ' ἐνέργειαν. Having done so, he identifies the active intellect of *de Anima* III.5, which is 'transcendent', 'changeless' and 'pure', with the intelligence of the *Metaphysics* Λ.[14] Finally, relying on the doctrine of the adventitious intellect of *de Generatione Animalium* (736b, 27ff.) he states that it comes into us from without. This adventitious intelligence is intelligible in the highest degree (κυρίως νοητόν), it is pure actuality, and the cause of other intelligibles.[15] Therefore it may be called 'productive'. It is true that in demonstrating his point Alexander contaminates the simile of the light of *de Anima* III.5, in which light is a condition for the visibility of colours, with that of the sun in the *Republic* (507e-509b), in which the visible par excellence provides visibility to other visible objects.[16] However, the productive aspect of the supreme intellect is not necessarily related to the idea of participation.[17] Since the productive intellect is intelligible in the full sense of the word, it is the cause of the actualisation of the material intellect insofar as it provides it with pure intelligibles in order for it to develop its highest activity.[18]

Alexander's differentiation between material and immaterial intellect correlates with his distinction between forms in matter (ἔνυλα εἴδη) and forms without matter (ἄυλα εἴδη). Whereas the material intellect can only perceive intelligibles in the presence of sensibles, the intellect *in habitu* when fully actualised can perceive them without the aid of sensorial perception.[19] The productive intelligence in turn only thinks immaterial intelligibles.

Of prime importance for the transformation of human into divine intelligence is Alexander's restatement of the Aristotelian theory concerning the identity of subject and object in the intellect's act of knowing things without

---

[12] Alex. Aphr., *de Anima* 85.11.

[13] Alex. Aphr., *de Anima* 85.25-86.5. See also 9.20ff.; 33.3ff.; 107.21ff. Cf. Arist., *de An.* 412a 9-11; 22-27 and *EN* 1157b 5ff.

[14] *De anima* 89.9-19.

[15] Alexander, *de Anima* 90.19ff.

[16] This is obvious in his statement that the supreme and highest good is the cause of goodness in other things (*De anima* 89.2-4).

[17] So Moraux, *Alexandre d'Aphrodise*, p. 39f.; Merlan, *Monopsychism*, p. 39; Donini, *Tre Studi*, p. 45, note 107; Fotinis, *The De Anima*, pp. 326-328.

[18] Lloyd, 'The Principle', p. 150; Schroeder, 'The Analogy', pp. 215-225.

[19] Alex. Aphr., *de Anima* 85.20-26.

matter.[20] Before this act, the knowing faculty and the thing known stand apart and are opposed as members of a relationship. When they are actualised their opposition disappears and they become a single reality.[21] When the intellect thinks the intelligibles it becomes itself an intelligible.[22] Consequently, when our intellect thinks the productive intelligence, it becomes in some way (πῶς γίνεται) this supreme intellect, and this is possible since in the act of thinking our intellect is assimilated (ὁμοιωθῆναι)[23] to the thing thought when apprehending its form. It follows that if one desires to have something divine within himself, he must engage in thinking something of this order.[24]

The goal of man's life seems to consist in rising from the lowest levels of materiality to the highest of intelligibility. The culmination of this process, which is basically rational and strictly related to cognition, brings about the reunion with the productive intellect by means of the fusion of subject and object in the act of knowing.

Other authors

It is very plausible that the notion of the adventitious intellect would already have been discussed at the end of the first century BC.[25] Given the fact that Philo, in the first century AD, mentions it a couple of times,[26] it is not surprising to find it combined with *de Anima* III. 4-5 in his philosophical allegory concerning the celestial and terrestrial man. After explaining how the mortal *nous* is mixed with the body, Philo explicitly states that the mind of man is capable of turning to and knowing God only thanks to the *pneuma* inbreathed by God himself.[27] Just as in Alexander, by being extrinsically

---

[20] Aristotle, *de An.* 430a 2-4; 431a 1-3; *Metaph.* 1072b 20-22; 1074b 38-1075a 5.

[21] Alex. Aphr., *de Anima* 86.23-28; see note 20 and further Plotinus, *Enn.* 5.9.5.30-31.

[22] Alex. Aphr., *de Anima* 86.14-23.

[23] Merlan (*Monopsychism*, p. 16) sees in the use of the term ὁμοιοῦσθαι (90.17) an indication for a possible Peripatetic version of the Platonic ὁμοίωσις θεῷ. Against this, cf. Donini, *Tre studi*, p. 36ff.

[24] Alex. Aphr., *de Anima* 91.5-6. Cf. Aristotle, *EN* 1177b 31-34. According to Donini (*Tre studi*, p. 36ff.), the religious tone in the former passage would be less important than in the latter. However, cf. Sharpless, 'Alexander', p. 1225, note 236.

[25] Aetius IV.5.11, 392b 3 Diels.

[26] *Opif.* 67 and *De Somniis* 30-32.

[27] *Leg. alleg.* 31-40; 88-90. See also *Plant.* 19; *Heres.* 64. See Moraux, *Der Aristotelismus*, pp. 408-409.

acted upon the material and perishable *nous* becomes immortal and divine *nous*.[28]

Alcinous is a clear example of the integration of Aristotelian noetics and psychology. He distinguishes both between embodied and disembodied intelligibles and between divine and human thought. Whereas God thinks νοητά or 'intelligibles' in their purity, i.e. Platonic ideas, man thinks exclusively forms-in-matter, in the Aristotelian sense, since his thinking is mixed with sensorial perception. As has been pointed out,[29] this idea may arise from Aristotle's assertion that man cannot think without images.[30] There is, however, a more proper precedent for this idea. It is Aristotle's statement that *intelligibilia*, though more knowable by nature (γνωριμώτερα τῇ φύσει), are less knowable than the particulars due to sensorial perception.[31]

Aristotelian elements also pervade Alcinous's exposition of the gradation soul – intellect in potentiality, intellect in actuality – first principle.[32] Potentiality and actuality of the intellect,[33] movement of the cosmic intellect due to the attraction to the object of desire,[34] God conceived of as 'unmoved mover' and intellect, and the identity of this highest intellect with *intelligibilia*[35] are clearly Aristotelian. For our present purpose, however, it is especially interesting to focus on the double causality of God,[36] namely final and

---

[28] For the influence of the Aristotelian 'unmoved mover' on Philo, see Boyancé, 'Le Dieu très haut'. For the idea of the reunion of subject and object in the act of knowing, see Wolfson, *Philo*, p. 229ff. and p. 249ff.

[29] Moraux, *Der Aristotelismus*, pp. 459-460.

[30] See *de An.* 431a 16-17; *Mem.* 449b 30-450a 14.

[31] See *Metaph.* 1029b. 3ff and *APo.* 71b. 33-72a 6; *Ph.* 184a. 15ff. Cf. also Max. Tyr. 11.7 (ed. Trapp). See Festugière, *La Révélation*, p. 111.

[32] *Didasc.* 164.18-20. It is true that in dividing the Intellect into two, namely first principle and celestial intellect, he is combining Aristotle's divinity with the Platonic good, but it is remarkable that this first principle is nevertheless described as an 'unmoved mover'. As Moraux (*Der Aristotelismus*, p. 467) suggests, this differentiation between a highest intellect and celestial intellect might proceed from combining the God from the *Metaphysics* Λ and the tendency to consider the celestial bodies endowed with a soul. It is indeed plausible that the idea of a celestial intellect proceeds from Peripatetic discussions concerning the movement of the celestial bodies, since Simplicius (*in Coel.* 379.32ff.) preserves a passage of Alexander's commentary on *Coel.* (ad II 1) in which the latter comments on contradictions between Aristotle's *Coel.* I 1-2 and II 2; 12ff. In this context, he refers to Herminus's opinion, according to which the celestial bodies could have an eternal movement thanks to their being endowed with a soul. Cf. Aristotle, *Coel.* 285a 29; 295a 18ff; Theophr. *Metaph.* 8.5b.2; Aetius I.7.32.

[33] *Didasc.* 164.18-20. Cf. *de An.* III. 5.

[34] *Didasc.* 164.16; 20-24. Cf. *Metaph.* 1072a 23-30.

[35] *Didasc.* 163.30-31. Cf. *Metaph.* 1072b 18-26; 1074b 17-35.

[36] Or 'supreme intelligence'. For Alcinous's hesitation concerning the first principle

efficient,[37] in turning the cosmic soul to himself. On the one hand, he moves the cosmic intellect attracting him as ὀρεκτόν.[38] On the other, he 'orders' (κοσμέω) the cosmic soul so that it might be attracted to the νοητά. According to Alcinous this 'ordering' should not be understood as 'creation' in time.[39] Rather, this action of the first God awakens the cosmic soul from a deep sleep[40] in order for it to be able to turn (ἐπιστρέφω) to him.[41] By thinking the thoughts of God,[42] the cosmic soul becomes a νοῦς κατ' ἐνέργειαν, that is the cosmic *nous*.[43]

Although Alcinous applies it not to the human but to the cosmic soul, this latter sort of causality presents obvious similarities with Alexander's. In addition, we see once again,[44] that in the process of actualisation of the *nous* there appears a contamination of *de Anima* III.4-5 and *de Anima.* II.1. From the former he takes the notions of potentiality and actuality of the *nous* and the definition of the 'intellect in act' as ὁ νοῦς τῷ πάντα ποιεῖν.[45] From the latter he takes the differentiation of two sorts of actuality: whereas the active exercise of knowledge (θεωρεῖν) is compared to awakening, the mere

---

(*Didasc.* 164.23: God; 164.26: intellect; 164.27: intellect; 164.31: God) see Merlan, *Monopsychism*, pp. 63-67.

[37] See Loenen, 'Albinus' Metaphysics (I), p. 302ff. As Moraux (*Der Aristotelismus*, p. 464) points out, although from an Aristotelian point of view, God's causality is that of a mover, Alcinous's main interest is not as much the cause of motion as the relationship of this world to the world of the ideas.

[38] *Didasc.* 164.24ff.

[39] *Didasc.* 169.31ff.

[40] Witt (*Albinus*, pp. 131-132) thinks the origin of the motif is unknown. However, by comparing it with Plutarch (*De anima. Procr.* 1026 F) and with Maximus of Tyre (10.1-3 Trapp), he concludes that it must have been a current notion in Middle-Platonic discussions on the soul. According to Loenen ('Albinus' Metaphysics II', p. 51), Alcinous is reworking a Plutarchean image. Trapp (*Maximus of Tyre*, p. 87) relates it to the Platonic passage *Meno* 85c. Cf. Philo, *De Somniis.* 165. See, however, *infra*, notes 46 and 50.

[41] *Didasc.* 165.1-4; 169.30-35.

[42] The Aristotelian definition of the 'unmoved mover' as νόησις νοήσεως in *Metaph.* Λ 6-9 and the doctrine of the identity of intelligibles and intelligence (*de An.* 430a 3-4; 19-22; 431a 1-4; *Metaph.* 1072b 18-22; 1074b 17-1075a 5) provides a suitable conceptual frame for the fusion of Platonic and Aristotelian thought in the doctrine that the ideas are thoughts of God. See Jones, 'The Ideas', p. 323; Rich, 'The Platonic Ideas'.

[43] *Didasc.* 164.27-31.

[44] See *supra*, note 15.

[45] According to Merlan (*Monopsychism*, p. 10) it is precisely the use of the verb ποιεῖν the link permitting Plotinus the identification of the Platonic demiurge (*Tim.* 28C, ποιητὴς τοῦ παντός) with the Aristotelian Intelligence.

possession of knowledge is equalled to sleep.[46] This double actuality provides Alcinous with a proper frame with which to sustain his allegorical interpretation of the *Timaeus*. Plato did not mean to say that the world was created in time, since it has always existed. Since a transition from a sleeping to a waking state does not involve qualitative change,[47] he avoids the contradiction of saying that the cosmic *nous* was in a potential state, while little before he described it as being in actuality (ὁ κατ' ἐνέργειαν πάντα νοῶν καὶ ἅμα καὶ ἀεί).[48] It must be noted that Alcinous describes the result of the awakening with the verb ἀποβλέπω 'to look at, contemplate',[49] which might echo the θεωρεῖν of the Aristotelian passage.[50] Only after being awakened can the cosmic soul contemplate the νοητά and be attracted to them.

A similar conception appears in Maximus of Tyre,[51] but now applied to the individual's intellect. He distinguishes human intellect, which also appears in its double aspect of potentiality and actuality, from divine intellect, which is defined as 'the intellect which thinks always all things at the same time'.[52] How does man come to know this intellect? Of importance is his

---

46 Although sleeping and waking are compared in the *Metaphysics* (1048a 30-1048b2) with potentiality and actuality, respectively, the simile is applied in *de An.* 412a 9-11; 22-26 to the differentiation between two sorts of ἐντελέχεια, namely the possession of knowledge (ἐπιστήμη) from the active exercise thereof (θεωρεῖν). Similarly, the simile of sleeping/waking distinguishes in *EN* 1157b 5-11 the disposition (ἕξις) to be virtuous from the active exercise of virtue. The virtue of a sleeping virtuous man is indeed a ἕξις by the fact that he is not exercising it (*EN* 1095b 30ff). Cf., however, *MM* 1201b 15ff; D. L. 5.34.

47 See Aristotle, *de An.* 417a 30-417b 2.

48 See in addition Aristotle's statement (*de An.* 431a 1-6) that potentiality is prior in time to actuality only in the particular cases but not in the universal.

49 *Didasc.* 169.40.

50 Deuse (*Untersuchungen*, pp. 83-85) agrees with Mansfeld ('Three notes', p. 65), in seeing here the influence of Aristotle's definition of sleep as 'inactivity of the soul' (*EE* 1219b 19ff.). However, in his view, the motif originates in Aristotle's differentiation between two kinds of potentiality in *de An.* 417b 29ff. (cf. *GA* 735a 9-11). The context of Alcinous's statement, however, seems rather to point to *de An.* II.1 (412a 9-11; 22-27), where sleeping and waking are considered as first and second actualities (see *supra*, note 45). In addition, the passage of Porphyrius (*ad Gaurum*, p. 33.14-23) quoted by Deuse (*Untersuchungen*, pp. 84-85) seems also to point in this direction (see especially 38.22-23). Also Dillon (*Alcinous*, p. 164), though hesitatingly, refers to *de An.* II.1, since in his view the use of ἕξις and ἐνέργεια would come into the Peripatetic and Platonic tradition later. See, however, *EN* 1157b 5-11.

51 All references follow Trapp, *Maximus Tyrius*; English translations by Trapp (*Maximus of Tyre*).

52 *Diss.* 11.8, 187-191.

differentiation between 'intelligibles' and 'perceptibles'. Where-as the former, though more knowable by nature, are less knowable due to sensation, the latter are more familiar to us through daily experience.[53] Since God is an intelligible,[54] it is obvious that no sense perception can grasp him. According to the principle of 'like knows like' it is the 'noblest and purest and most intelligent' aspect of his soul that can grasp him at once in one single act of comprehension. In this intuitive and single act of apprehension the intellect 'sees' and 'hears' the intelligible.[55] In describing the method to achieve this intuitive apprehension, namely the ἐπιστροφή, *Oration* 11 combines this 'inward turn' with the topos of the flight of the mind and follows, consequently, the Platonic *Phaedrus* (246d-247e). However, Maximus applies it to describe the way in which the νοῦς, finally freed from sensation, achieves ultimate knowledge.[56] In addition, his treatment of the issue in *Oration* 10 presents interesting parallels with the above-mentioned conception of Alcinous. In order to answer the question of whether this 'turning to oneself' must be called 'learning' or 'recollection' he states the following, 'Reason (λόγος) ... does not bring or implant understanding (ἐπιστήμη), like something the soul did not already possess; instead, it reawakens the understanding it does possess, but which is dim, and constrained and torpid'.[57] Knowledge is the 'awakening' and the 'organisation' of true opinions.[58] We find then once again the idea of the double actuality and, together with it, the notions of awakening and ordering. Whereas 'forgetfulness' is comparable to sleep, the process of rousing provoked by reason is 'recollection' and the preservation of what has been set in order is 'memory'.[59]

The Acts of Andrew

In the *Acts of Andrew*[60] (= *AA*) the notions of the awakening of human intellect and the active exercise of knowledge are also preconditions for the possibility of returning to God (ἐπιστρέφω). Until God's intervention the

---

[53] *Diss.* 11.7, 132-134. *Supra*, note 31.
[54] For the slide from God as supreme intelligible (νοητόν) to God as supreme intelligence (νοῦς), see Festugière, *La Révélation*, pp. 112-113; 127; 136 and Trapp, *Maximus of Tyre*, p. 103, note 32. See also *supra*, note 42.
[55] *Diss.* 11.9, 204-213.
[56] See [Aristotle], *Mu.* 391b 8-11; Philo *Opif.* 69-71.
[57] *Diss.* 10.3, 73-6. See *supra*, notes 40, 46, and 50.
[58] *Diss.* 10.6, 142-3. Cf. Aristotle, *APo.* 100a 3-100b5.
[59] *Diss.* 10.6, 159-61.
[60] Bonnet, 'Ex actis Andreae', in *AAA* II/1, pp. 38-45. All English translations are my own.

σῳζόμενον γένος or 'saved race' remains ignorant of its true condition and is apparently subject to time and movement and the concomitant processes of coming-to-be and passing away. By recognising his people, however, God confers upon them a *new* existential status.[61] Awakened from the slumber of oblivion (λήθη),[62] their disposition to know allows them first to acknowledge their true belonging and to reach dialectically —namely, by the *via eminentiae*— knowledge of God's existence.[63] Even if just preparatory, this preliminary knowledge is essential, since the awareness of their transcendent origin also provides them with the goal (τέλος) of their being. As they possess now something to turn themselves to, they can engage in the process of self-knowledge that, by means of introversion, will allow them to come to know God essentially.

The lower stages of this process of self-knowledge are placed within the frame of the known Aristotelian triad for acquiring moral virtue,[64] namely φύσις, ἔθος-ἄσκησις, and μάθησις, that frequently appears in philosophical writers of the period.[65] Similarly as in Aristotle, these three successive steps strictly correspond with the domain of practical life and φρόνησις as opposed to theoretical life. Although the former is clearly inferior to the latter, man's actual 'unfree' state makes practical life a necessary pre-condition for contemplation.[66] This is the reason why the return to God is conceived as a deconstructive process. Before human intellect can achieve an intuitive apprehension of the divinity, it must be released from

---

[61] Bonnet, 'Ex actis Andreae', 38.5-10: Μακάριον ἡμῶν τὸ γένος, ὑπὸ τίνος ἠγάπηται; μακαρία ἡμῶν ἡ ὕπαρξις, ὑπὸ τίνος ἠλέηται; οὐκ ἐσμέν τινες χαμαιρριφεῖς, ὑπὸ τοιούτου ὕψους γνωρισθέντες· οὐκ ἐσμὲν χρόνου, εἶτα ὑπὸ χρόνου λυόμενοι· οὐκ ἐσμὲν κινήσεως τέχνη, πάλιν ὑφ' ἑαυτῆς ἀφανιζομένη, οὐδὲ γενέσεως αἰτία, εἰς αὐτὸ τελευτῶντες.

[62] See *Phaedr.* 248C; 250A, but also *C.H.* X 15; *TripTrac* 77.22-25; *EvVer.* 17.10ff.

[63] Bonnet, 'Ex actis Andreae', 38.11-18: ἐσμὲν τοῦ κρείττονος· διὰ τοῦτο ἀπὸ τοῦ χείρονος φεύγομεν. ἐσμὲν τοῦ καλοῦ, δι' ὃν τὸ αἰσχρὸν ἀπωθούμεθα (-μεθα prop.Bonnet)· τοῦ δικαίου, δι' οὗ τὸ ἄδικον ῥίπτομεν· τοῦ ἐλεήμονος, δι' οὗ τὸν ἀνελεήμονα ἀφίεμεν· τοῦ σῴζοντος, δι' οὗ τὸν ἀπολλύντα ἐγνωρίσαμεν· τοῦ φωτός, δι' οὗ τὸ σκότος ἐρρίψαμεν· τοῦ ἑνός, δι' οὗ τὰ πολλὰ ἀπεστράμμεθα· τοῦ ὑπερουρανίου, δι' οὗ τὰ ἐπίγεια ἐμάθομεν· τοῦ μένοντος, δι' οὗ τὰ μένοντα εἴδομεν.

[64] See *EN* 1179b 20ff; *EE* 1214a 16ff.

[65] See [Plutarch], *De lib. ed.* 2a; Alcin., *Didasc.* 181.19-182.13; D.L. 5.18; Archytas, *De ed.* 3, p. 41.20ff. Thesleff; Philo, *Vita Abrah.* 52-4; Maximus of Tyre 1.5;12.9; 27.9. Arius Didymus (*ap.* Stob. *Anth* 2.118.5ff) offers perhaps the best parallel for *AA*, since his triad is φύσις, ἔθος and λόγος.

[66] See also Alcinous, *Didasc.* 153.3-23 and Aspasius, *in EN comm.* I, 2-2, 13. Cf. Aristotle. *EN* 1177b 7ff. See Bos, *Cosmic and Meta-Cosmic Theology*, pp. 151-152.

the burdens that it acquired in its degradation from the highest to the lowest regions of the cosmos.

As for the cause that originates this process of degradation, *AA* is not explicit.[67] The first time we encounter the νοῦς in the key passage of the Paradise scene,[68] it has been already degraded to the level of ψυχή or 'soul', since it is said to be subject to affections, namely to 'suffering' (πάσχω).[69] However, the reason explaining its suffering describes the νοῦς as being 'split up and alienated from itself' (καταχθεὶς καὶ ἀπολισθήσας ἑαυτοῦ).[70] This may give us a hint as to the cause of its degradation, since it actually depicts the loss of the intrinsic characteristics of the νοῦς, namely its unity and self-centred activity.[71] It is thus plausible to think that what originates the degradation of the νοῦς is the incorrect direction of its intellective activity. An interruption (due to necessity related to creation?) in the intellect's incessant activity of thinking the intelligibles, namely God, might originate a first discrepancy between subject and object in the act of knowing. This possibility is coherent with the reasons given for the degradation of Eve and Adam, namely ignorance and error. The deficiency of the νοῦς and its subsequent intermittent thinking combined with Eve's and Adam's unawareness of this deficiency, might have led them to try to supersede their ignorance by their own means.[72]

This interruption in the intellect's incessant activity originates three causally related steps in a gradual withdrawal of the 'intellect' from its original, complete and unitary knowledge. The first one is indeed the transition from intuitive to discursive apprehension due to the discrepancy between subject and object (and this also implies the appearance of number).[73]

---

[67] In spite of Erbetta (*Gli Apocrifi*, p. 400, note 1) Eve and Adam's fall in *AA* is normally interpreted as due to sexuality. See Hennecke, *Handbuch*, p. 552; Wagener, 'Repentant Eve, Perfected Adam', pp. 350-351; Cirillo, 'L'Uomo interiore', p. 15, and most recently Bovon, 'The Words', p. 84. For an important interpretative breakthrough, see Luttikhuizen, 'The Religious Message', pp. 99-100.

[68] Bonnet, 'Ex actis Andreae', 40.12-22.

[69] Bonnet, 'Ex actis Andreae', 40.15-17: καὶ ὅπερ ὁ σὺν ἐκείνῃ καταχθεὶς καὶ ἀπολισθήσας ἑαυτοῦ νοῦς ἔπαθεν, ἐγὼ σὺν σοί, τῇ γνωριζούσῃ ἑαυτὴν ἀναγομένην, διορθοῦμαι.

[70] Bonnet, 'Ex actis Andreae', 40.15-16. The first of the participles is normally considered to belong to the verb κατάγω '(metaph.) bring down'. I understand it as an aor. participle from κατάγνυμι 'break in pieces, shatter'.

[71] *Metaph.* 1072b 20-1073a 14; *de An.* 430a 22; Plot. *Enn* 5.9.5.

[72] This does not imply, as in *TripTrac.* 78.17, 'arrogance' (*tsise* or τόλμα, cf. Plot. *Enn.* 5.5.1), since *AA* explicitly relates the 'suffering' with ignorance (Bonnet, 'Ex actis Andreae', 40.13-15).

[73] For the νοῦς as ἓν πολλά see Dörrie, 'Zum Ursprung', p. 286.

And as soon as the act of knowing does not revert upon itself there appears ignorance. It is this very ignorance that responds for the second stage of degradation from νοῦς to soul or ψυχή, since it originates the appearance of doubt and fear, namely affections.[74] Desire to achieve knowledge in order to overcome ignorance impels the soul to undertake the cognitive process. Paradoxically, however, this search for knowledge will only enhance its disgrace. Since λόγος or 'discursive knowledge' is intrinsically related to reason and speech, its application to gain knowledge will give up soul and intellect to the slumber of appearance and persuasion.[75] As was to be expected, the original ignorance remains unaltered and characterises each of these stages of degradation.

This interpretation is further supported by Andrew's assertion that, differently from Eve and Adam, Maximilla and he himself are now aware of the first couple's deficiency. It is precisely this awareness of that which was defective in Eve (τὸ ἐκείνης ἐνδεές) and that which was imperfect in Adam (τὸ ἐκείνου ἀτελές) that allows Maximilla and Andrew to correct their error. Their being conscious of their true nature and origin enables them to acknowledge their deficient condition. This awareness is essential inasmuch as it generates the will[76] that will allow the ἐπιστροφή or 'return to God', since they now distinctly perceive themselves as the effect of a cause and they subsequently recognise their potential identity with it.[77]

If my interpretation of the passage is correct, Aristotelian echoes are evident not only in the conception of the νοῦς as thinking itself always and incessantly, but also in the notion that what is potential, incomplete and deficient tends to its perfection and actuality. The Aristotelian assertion that actuality is prior to potentiality[78] plays here, as it also does in Neoplatonism[79] and Gnosticism[80], an important role, since it is the full actuality or perfection of the cause which assures the eventual perfection of the effect.

---

[74] For a similar process in the origin of affections see *EvVer.* 17.10-15.

[75] Bonnet, 'Ex actis Andreae' 44.10-14: ὁρᾶτε γὰρ μηδὲν (Bonnet: prop. for μηδένα) ὑμῶν ἱστάμενον, ἀλλὰ τὰ πάντα εὐμετάβολα μέχρι ἠθῶν ἀνθρωπίνων· τοῦτο δὲ συμβαίνει διὰ τὴν ἀπαίδευτον ψυχὴν τὴν εἰς φύσιν πλανηθεῖσαν καὶ κατὰ ⟨τὰ⟩ (*scripsi*) τῆς πλάνης ἐνέχυρα κατέχουσαν.

[76] Plot. *Enn.* 6.1.21.

[77] Actuality is their ἀρχή and their τέλος. See Proclus, *Inst.* 31, and Dodds, *Proclus* (*ad loc*).

[78] See *de An.* 430a 20-22; *Metaph.* 1050a 7ff.; 1077a 18ff.

[79] Plot. *Enn.* 6.9.9. Cf. Proclus *in Parm.* 922.3ff.

[80] *EvVer.* 19.3-7. Attridge and MacRae (*Nag hammadi Codex I*, p. 53) provide several parallels, but none presents so explicit a treatment of the issue. *TripTrac.* 78.1-17 (cf. 123.32-33) presents certain similarities, although it relates 'imperfection' with 'arrogance', since according to this text the effect seems to

However, before reaching this level the followers must engage, under Andrew's supervision, in a laborious and complete deconstruction of their current being. In order to reverse the effects of the cognitive (discursive) process, the knowledge transmitted by Andrew must awaken soul and intellect from their present dormant state and organise the disorder arising from material life. Andrew explicitly states that his goal is 'to remind' (ὑπομιμνήσκω)[81] and we have seen that Maximus of Tyre compares 'forgetfulness' with sleep.[82] This awakening is *sensu stricto* possible since the intellect is extrinsically acted upon. 'To be recognised' (γνωρίζομαι) by the One assures, by the principle of 'like knows like', the possibility of knowing him (γνωρίζω).[83] The following rational process of deconstruction, which somewhat resembles the *via eminentiae*, intends to neutralise the fore-mentioned three successive steps that have led the intellect to its state of apathy. These reverted steps are represented by Andrew's speeches to the brethren, Stratocles and Maximilla[84] and all of them present clear Aristotelian traits. The first one (φύσις) consists in reorganising the realm of perception and being able to distinguish false from true appearances. The second one (ἔθος) consists in reinforcing the rational part of the soul that may subdue excessive affections by means of rational control and μετριοπάθεια.[85] The ultimate instruction, or third step, aims to supersede both speech and the discursive character of λόγος. Andrew's words of praise remark the success of the deconstructive stages:[86]

> Well done you, saved nature, since you were not strong and you did not keep yourself hidden![87] Well done you, soul, shouting which you suffered and re-

preserve the perfection of its cause. Cf. *TripTrac.* 69.1-4 and Thomassen, *Le Traité*, p. 344. Differently, *EvVer.* 19.3-7 states that the father 'retains within himself their perfection, granting it to them as a return to him and a perfectly unitary knowledge'. The awareness of their deficiency proceeds from self-knowledge (See *EvVer.* 21.3-25, esp. 14-18; cf. Bonnet, 'Ex actis Andreae' 40.17-19). Cf. Ménard, *L'Évangile*, p. 91.

[81] Bonnet, 'Ex actis Andreae', 44.4-8: Ἐγὼ ἀδελφοὶ ἐξεπέμφθην ὑπὸ τοῦ κυρίου ἀπόστολος εἰς τὰ κλίματα ταῦτα ὧν με κατηξίωσεν ὁ κύριός μου, διδάξαι μὲν οὐδένα, ὑπομνῆσαι δὲ πάντα τὸν συγγενῆ τῶν λόγων ἄνθρωπον, ὅτι ἐν κακοῖς τοῖς προσκαίροις διάγουσι, τερπόμενοι ταῖς ἐπιβλαβέσιν αὐτῶν φαντασίαις·

[82] Once he does so he advises his followers to stay awake (Bonnet, 'Ex actis Andreae' 45. 28: διυπνίζω). Cf. Aristotle, *de An.* 429a 5ff.

[83] Bonnet, 'Ex actis Andreae', 38.1-20.

[84] Bonnet, 'Ex actis Andreae', 44.4-45.33; 42.7-43.18; 39.27-42.6, respectively.

[85] Since τὸ πᾶν ἦθος διὰ ἔθος (Plat. *Legg.* 792e), see Aristotle, *EE* 1220a 39.

[86] Bonnet, 'Ex actis Andreae', 40.24-27.

[87] My translation of Schimmelpfenning's reading (in: Hennecke, *Handbuch*, pp.

turning to yourself! Well done Anthropos, who understands what is not yours and hastens towards what is yours!

Once so far, practical wisdom or φρόνησις can be left behind in order to achieve a higher degree of intuitive apprehension. It is at this point that the listener, namely the intellect, thinks much worthier things than what has been said.[88] By becoming aware of being 'immaterial, saint, light, akin to the not-generated, transparent, intelligible, celestial, and pure',[89] he perceives and apprehends himself in his real condition. He then sees his face in his own essence and tearing all his shackles he desires (ποθέω) to see the One whom he has not yet seen.[90] Thanks to God's activity as *causa efficiens* in providing the means for his people to turn to him, he attracts them now as *causa finalis*. In this reunion with God, which is equalled to the fusion of subject and object in the act of knowing, nothing obscure remains and by means of a sort of illumination man achieves complete knowledge.

One might rightly wonder whether this illumination resulting from intuitive apprehension may at all be compared with the τέλος of philosophy according to Aristotle. However, famous is the Aristotelian assertion that human nature is 'un-free', which is explicit in the *Metaphysics* and implicit in other places, such as the *Nichomachaean Ethics* or the *Physics*[91]. Aristotle's defence of the θεωρητικὸς βίος or 'contemplative life' as the most excellent for man is consistent with this view. Such a life would be his complete *eudaimonia*, since it is the activity in accordance with the highest virtue of his highest or most divine part, namely his intellect.[92] However, as humans also depend on externalities and are ephemeral, man cannot achieve this absolutely by virtue of his human condition, but rather by virtue of participating in what is most divine in him. The corollary of the preceding statement is that mortals should try 'to achieve immortality as far as possible'.[93] True, the method to reach this goal remains obscure and perhaps the *Eudemian Ethics* advances a little by putting in the contemplation of God the object of this contemplative life.[94]

552-553): μὴ ἰσχύσασα μηδ' ἑαυτὴν ἀποκρύψασα.

[88] Bonnet, 'Ex actis Andreae', 40.27-28, with Festugière's emendation (*La Révélation*, p. 229, note 3): νοούμενον ἢ ⟨τὸ⟩ λεγόμενον.

[89] Bonnet, 'Ex actis Andreae', 40.32-3.

[90] Bonnet, 'Ex actis Andreae', 41.2-6.

[91] In general, see *Metaph.* 982b 22-983a 5; for human need of external goods as a condition to contemplation, see *EN* 1178a 33-35; for man's dependence upon the environment, see *Ph.* 253a 12ff; 259b 11.

[92] *EN* 1177a 12-13; 21-22.

[93] *EN* 1177b 34-35.

[94] *EE* 1249b 16ff.

One might, indeed, be tempted to relate these notions to the passage in *de Anima* that states that θεωρητικὴ ἐπιστήμη or 'speculative knowledge' is the same as its object. Since in the case of things without matter that which thinks and that which is thought are the same, there seems to be only one step to proposing that the final goal of the cognitive process and the reunion with God are identical. Since the goal of man is to live in accordance with his most divine part, and since this part is his intellect, understood not as potentiality but as an act, it would seem, indeed, that at those moments in which human intellect is actualised, man, by the act of thinking God, is assimilated to him. Such an act of intelligence might be seen as a return to a perfect and unitary knowledge that transcends, after the reunion of subject and object, the very discursive knowledge, by reaching a sort of supra rational, intuitive apprehension.

As far we know, it is true, Aristotle did not explicitly make this step and the resulting ambiguity is what Zeller calls the 'mystische Unbestimmtheit' of his philosophy.[95] Significantly, however, the corpus repeatedly draws the distinction between intuition of the intellect and the discursive λόγος and describes the former with the words θιγεῖν, θιγγάνειν, θῖξις 'to touch, contact, contacting'.[96] Two Aristotelian fragments advance in the definition of this contacting. The first one, *De philos.* 15R preserved by Michael Psellus,[97] after referring to the same distinction states that intuitive apprehension comes about by means of a sort of illumination of the *nous* (αὐτοῦ παθόντος τοῦ νοῦ τὴν ἔλλαμψιν). In the second one, *Eudemus* 10R preserved by Plutarch,[98] Aristotle is said to have conceived the culmination of philosophy as a sort of telestic illumination, by which the soul 'contacts' (θιγγάνειν) the first, pure and simple reality.[99]

On the basis of the closing remarks to the *Posterior Analytics*, one may suggest that for Aristotle the cognitive process, although its lower stages seem to be empirical, culminates in a non-discursive act of intelligence.[100] Of course it might be objected that intuitive apprehension of the principles does not necessarily mean supra-rational apprehension. We seem then to face the same aporia that E. Zeller faced more than a hundred years ago

---

[95] Zeller, *Die Philosophie*, p. 793.

[96] As pointed out by Theiler ('Die Entdeckung', p. 105) and Merlan (*From Platonism to Neoplatonism*, p. 186, and *Monopsychism*, p. 3). See also Bos, *Cosmic and Meta-Cosmic Theology*, pp. 64-65. Cf. *Metaph.* 1051b 24; 1072b 19-21; *de An.* 427b 4. Furthermore Plotinus, *Enn.* 5.3.10.39.

[97] *Schol. ad Joh. Climacum* 6.171.

[98] Plutarch, *De Is. et Os.* 382d-e.

[99] See Boyancé, 'Sur les mystères d'Eleusis', p. 462; Croissant, *Aristote et les mystères*, pp. 145-146.

[100] *An. Post.* 100b 5ff.; *EN* 1140b 31-1141a 8.

when interpreting the term ἀρχαί. Either the principles are to be 'logical principles of reasoning', in which case we will have a rationalistic Aristotle, or they must be understood as 'principals', namely intelligibles, in which case our philosopher turns out to be supra-rationalistic.[101]

This survey of Aristotelian influences in different contexts of Late Antiquity shows that such an aporia did not exist for his ancient readers. As a point of fact the latter always take the Philosopher in the latter way. It is up to us to decide whether this interpretation is legitimate or not. In my view, however, this is perhaps less important and certainly less interesting than understanding how and why this *Aristoteles mysticus* develops in the period.

---

[101] See Zeller, *Die Philosophie*, p. 196; Merlan, *Monopsychism*, p. 31.

TALMUDIC TRADITION IN A CHANGING SOCIETY

Wout van Bekkum

The rabbis of late Antiquity and the early Middle Ages are not a phenomenon in Jewish tradition to be compared with mullahs in Islam, let alone reverends and priests in Christianity. Their spiritual or clerical attitude is but an aspect of the totality of religious observance, wonderfully summarised in a saying from the first century:

> I am a creature of God, my neighbour is also a creature of God; my work is in the city, his work is in the field; I rise early to my work, he rises early to his. Just as he does not presume to do my work, so I do not presume to do his work. Perhaps you say: I do great things and he does small things! We have learnt: It matters not whether one does much or little, if only he directs his heart to heaven.[1]

Early rabbinism had to define a way of life which was not exclusively connected to the formalities of Jewish ritual at the Temple of Jerusalem. The pursuit of learning, *Torah* in Hebrew, was one of the highest goals although its main function was not imposing a discipline but keeping Judaism practically and spiritually perceptive.[2] Influenced as it was by Christian judgment throughout the centuries it is a distortion to say that Judaism consisted exclusively of performing rituals or moral deeds. In rabbinic literature man is considered as more than what he does, and what he does is spiritually a minimum of what he is. Judaism therefore is not a purely biblical religion, and the rabbis consider themselves as the transmitters of new principles and ideas. Without their interpretation the biblical text is often unintelligible, thus rabbinic Judaism is based upon a minimum of revelation and a maximum of interpretation, in their own words, upon the will of God and upon the understanding of Israel. This understanding of the biblical word was not cheaply or idyllically won. The basic concept is the idea of 'oral teaching', that is to say, the supposed revelation of words and texts at the mountain of

---

[1] Babylonian Talmud *Berakhot* 17a: 'A favourite saying of the rabbis of Yavneh'.

[2] For a general overview, see Stemberger, *Introduction*; Reif, *Judaism*.

Sinai includes Scripture and an equivalent complex of traditions generated by oral exposition. A talmudic tradition summarises this idea in a story:

> A certain heathen once came before Shammai and asked him, 'How many Torahs have you?' 'Two', he replied: 'the written Torah and the oral Torah'. 'I believe you with respect to the written, but not with respect to the oral Torah; make me a proselyte on condition that you teach me the written Torah only'. Shammai scolded and repulsed him in anger. When he went before Hillel, he accepted him as a proselyte. On the first day he taught the Hebrew alphabet: *aleph*, *beth*, *gimmel*, *daleth*, etc. The following day he reversed them to him. 'But yesterday you did not teach them to me thus', he protested. [Hillel said:] 'Must you then not rely upon me? Then rely upon me with respect to the oral Torah too'.[3]

It is undeniable that such an oral tradition of explanations alongside Scripture must already have existed in the biblical period, and even more so at a later time. There are numerous rabbinic attestations of the oral mode of transmission of instructions and stories, and there is explicit testimony of written texts of legislative or narrative character. However, there are extreme difficulties in determining the aspects of the transition from oral tradition to the private and public use of written texts; it is at any rate important to know that talmudic discussions in their written form imply a certain mobility and actuality which certainly contributed to the almost exclusively literary approach commonly used by scholars until recent days. Contemporary history and empirical study were largely ignored, and the texts thereby came peculiarly timeless.

During the last fifty years the study of Late Antique Palestine and Babylonia has been greatly advanced. Modern interpretation of rabbinic texts increasingly considers historical aspects, although the texts themselves never transmit dates and hardly give an opportunity to correlate biographical details of the individual sage with facts known from secular history. The rabbi who was always considered as a sage and a teacher for his own generation and the generations to come is now treated in connection with concrete historical reality.[4] This approach, however, is not void of danger: some scholars tend to romanticise both the sages and their world, viewing them as having been spiritual leaders whose opinions directly influenced and largedly shaped Jewish society as a whole during a long period. Others cast grave doubts upon the centrality of the rabbis, and some adopt an ex-

---

[3] Babylonian Talmud *Shabbat*, fol. 31a.

[4] Feldman, 'Torah and Secular Culture'; Stemberger, *Jews and Christians*.

treme position wherein any influence they might have had over the people is greatly minimised.

The need to understand the status and position of these teachers in the talmudic society becomes more important than ever.[5] First of all, Jewish life was very complex and heterogeneous, as is evidenced from the dictum of rabbi Yohanan: 'Israel did not go into exile until they were splintered into twenty-four heretical sects'.[6] The rabbis faced a clear challenge: how to unify the people and to reach some level of consensus. For that reason one can regard the position of legislative and judicial bodies like the local court houses and the high court or Sanhedrin as instructive. After the destruction of the Temple in Jerusalem the high court was transferred to Tiberias in Galilee but the rabbis had a figurative way of saying that at Tiberias the authority of the Sanhedrin had sunk to its lowest level, Tiberias being a city on Lake Galilee below sea level.[7] Nevertheless, an entire talmudic tractate was devoted to the different types of Sanhedrin which had existed during the Temple period. When the first collection of legislative rules concerning the court system in Temple times was compiled towards the end of the second century, this was something of the past. The composition of the courts is discussed, enumerating the cases, civil, criminal, religious or political, which were brought before a court of three judges. Another section discusses the rights of the parties to a suit to choose or reject judges in courts of arbitration, as well as the rejection of witnesses. The grounds on which judges or witnesses can be disqualified are given *in extenso*. Procedures in civil and capital cases are differentiated and rules for the cross-examination of witnesses are offered. A large set of rules regards cases of discrepancies or contradictions in the evidence, the mode of procedure on the part of the judges at the voting and at the promulgation of the sentence, and the sentence practices.[8]

There can be no doubt that major political and religious forces within Jewish society such as the Jewish patriarchs or ethnarchs in Palestine (until the fifth century) were aware of this rabbinic-juridical compilation and employed rabbis in their service who would take on office at the patriarch's court, which represented the highest court of appeal in Jewish law.[9] The talmudic rabbis were granted appointments as judges, thus showing their activities in Jewish communal affairs, and in their exegesis of the legislative

---

[5] Urbach, 'Class-Status and Leadership', pp. 1-37; Berger, *Rabbinic Authority*, pp. 75-103.
[6] Palestinian Talmud *Sanhedrin*, fol. 10,29c.
[7] Babylonian Talmud *Rosh Hashanah*, fols. 31a-b.
[8] Brody and Herr, *Ephraim E. Urbach*, pp. 28-42.
[9] Horsley, *Archeology, History, and Society*.

tradition of the Sanhedrin they succeeded in nuancing and modifying its meaning in two directions for which I should like to adduce two examples. Firstly, the famous verse 'eye for eye' is discussed as follows:

> The Divine Law says 'eye for eye' – why not take this literally to mean putting out the eye of the offender? Let not this enter your mind, since it has been taught: You might think that where he put out his eye, the offender's eye should be put out, or where he cut off his arm, the offender's arm should be cut off, or again where he broke his leg, the offender's leg should be broken. Not so, for it is laid down, 'If anyone takes the life of a human being ... Anyone who takes the life of someone's animal':[10] just as in the case of killing a beast restitution is to be paid, so also in the case of killing a restitution is to be paid. ... How do you know that it refers to pecuniary restitution? Why not say that it really means capital punishment? Let not this enter your mind; first, because it is compared to the case dealt with in the text: 'Whoever kills an animal must make restitution', and furthermore, because it is written soon after, 'whatever he has done must be done to him', thus proving that it means pecuniary restitution.[11]

What is so special about this tradition? It is the strict application of the rhetorical device of analogy (literally in Hebrew *gezerah shawah*, 'parallel decree', in Greek *synkrisis pros ison*) that leads to this remarkable conclusion. No wonder that one of the rabbis could add to the argument that 'eye for eye' means pecuniary restitution: 'Because what then will you say when actual retaliation is meant, and the eye of one was big and the eye of the other was little, for how can in this case the principle of eye for eye be applied? The Divine Law states that you shall have one manner of law, implying that the manner of law should be the same in all cases'.

Exegetical adaptation is one way of actualisation, storytelling is another: by means of the same rhetorical device of analogy the verse from Daniel 4.14, 'The decision is announced by watchers, and the holy ones declare the verdict', is the point of departure for a large number of stories about the heavenly tribunal, consisting of God and his so-called *pamelia* (from Latin: *familia*), that is, the angels.[12] The description of scenes from the celestial court may be taken as an apocalyptic vision, but God's justice is used in some stories as an allegory for the history and future of the relationship between Israel and the nations:

---

[10] The discussion is based on Lev. 24, 17-19.

[11] Babylonian Talmud *Baba Qama* 83b-84a (Soncino edition, London, 1935, pp. 474-480).

[12] Babylonian Talmud, *Sanhedrin*, fol. 38b.

> In times to come, the Holy One, blessed be He, will take a scroll of the Law in His embrace and proclaim: 'Let him who has occupied himself herewith, come and take his reward'. Thereupon all the nations will crowd together in confusion. The Holy One, blessed be He, will then say to them: 'Come not before Me in confusion, but let each nation come in with its scribes'. Thereupon the kingdom of Edom (the Roman Empire) will enter first before Him. He will then say to them: 'Wherewith have you occupied yourselves?' They will reply: 'O Lord of the Universe, we have established many market-places, we have erected many baths, we have accumulated much gold and silver, and all we did this for the sake of Israel, that they might have leisure for occupying themselves with the study of the Torah'. He will say in reply: 'You foolish ones among peoples, all that which you have done, you have only done to satisfy your own desires. You have established market-places to place courtesans therein; baths, to revel in them; as to silver and gold, that is Mine, as it is written: 'Mine is the silver and Mine is the gold, says the Lord of Hosts' (Haggai 2.8). They will then depart crushed in spirit. Many other nations will step forth, and they will say: 'The others have oppressed Israel, but we have not' ... All of them will be judged including the people of Israel, because none accepted the Divine Law, until He came to Israel, suspended the mountain of Sinai over Israel like a vault, and said unto them: 'If you accept the Torah, it will be well with you, but if not, here you will find your grave'.[13]

This very interesting narrative counters any idea of divine election and demonstrates that the rabbis were not prepared to exercise active resistance despite all the generally traditional complaints about their subjection to the foreign Roman-Christian-Byzantine rule.[14] These masters pin all their hopes on salvation brought about directly by God at the end of time. Statements about the government occasionally show by their terminology that they reflect a historical period, but otherwise do not differ from earlier statements. No single rabbinic expression, taken by itself, can be understood as an indisputably proven reaction to Christianity or Islam, to Byzantines or Arabs, as the new powers. The historical colouring in the Talmud therefore does not arise from the mention of historical persons and events, but from the language of daily life, and the titles, institutions, offices, and bureaucratic procedures mentioned which stamped the Roman-Byzantine empire in the West or the Persian, Parthian and Islamic rule in the East.[15] The rabbis seem to give the impression of having separated themselves as far as possible

---

[13] Babylonian Talmud *'Avodah Zarah*, fols. 2b-3a (Soncino edition, pp. 2-3).
[14] Feldman, 'Some Observations'.
[15] Schäfer, *The Talmud Yerushalmi*.

from the pervasive influence of the non-Jewish surroundings. They attempted to retain their own rhythm of life and their attitude brought a pronounced quietism, but this could well be defined as a deliberate self-deception, theirs and ours. The Jews of Palestine received their legal and state support for most of the time under Byzantine rule, as one can learn from the legislation issued by Constantine, Theodosius and Justinian, and they were declared 'the tolerable people of the Book' after the Arab conquest. The privileges of the rabbis in the Jewish community and the relative freedom of worship and religion in Gentile society cannot have failed to have an impact upon the intentions of rabbinic teachings.[16]

In the context of the talmudic period both Greek and biblical Hebrew stood for a rich cultural content, each in its own way. Aramaic, and later Arabic, were the means of communication and intruded into the realm of religious literature. Unlike the first centuries CE, the talmudic period reveals an increasing interest on the part of the rabbis in civil and agricultural matters, although their main focus was on their schools and academies. Gradually they emerged as a new but unofficial social infrastructure in the Jewish community of many cities in Palestine and Babylonia. Notwithstanding the strong bonds which united these rabbinic centres, their commitment to the study of the Bible and to its fulfilment in religious observance, there were many differences among them. In cities like Caesarea, Sepphoris and Tiberias, a number of sages adopted a cosmopolitan posture by their knowledge of Greek and their familiarity with Greek literature. Such attitudes were also present in Babylonia where the issue of rabbinic authority over the people's life is even harder to define.[17] The talmudic sources do not convey a true and complete picture of how the laws shaped daily affairs, because they mostly suppress, or report only by indirection, actions and conduct contrary to rabbinic dicta. Archeological evidence, however, provided the most striking examples: the discovery in 1932 of the lavishly decorated synagogue of Dura-Europos in Syria with its mural paintings of persons and animals cannot be supported by any rabbinic interpretation of synagogue art. Hebrew and Aramaic incantation texts from Babylonia tell us considerably more about the attitudes of the common people than does the Talmud. What they tell us is that the people did not hesitate to make use of the magical science of their day in achieving their life-purposes. Since rabbinic homilies often refer to obedience to the dietary laws and presence at the ritual baths, then one would have to suppose that people did not obey the dietary laws and did not attend ritual baths. Simi-

[16] Rabello, *The Jews in the Roman Empire*; Levine, *The Ancient Synagogue*.
[17] Levine, *The Rabbinic Class*.

larly, the sages themselves preserved comments on the mores of their contemporaries which present a picture quite at variance from the one we should derive from strictly legal sources: 'Rabbi Giddal said in the name of Rav: If an inhabitant of Naresh kisses you then count your teeth. If a man of Nehar Pakod accompanies you, then it is because of the fine garments he sees on you. If a man of Pumbeditha accompanies you, change your lodging [that he may not rob you]'.[18]

We have to accept that any assessment of talmudic tradition in a multifaceted and rapidly changing society leads to different reflections: on the one hand the rabbis originated from the people and were involved in its life; on the other, as a religio-social elite, their beliefs, attitudes and way of life were not always acceptable to or identical with the people. In the talmudic period, roughly the third to seventh centuries, the rabbis did not control any communal institutions through which they could force their views upon the masses. This can be asserted to some extent when one studies the growing amount of evidence reflecting Jewish society in Palestine and Babylonia: apocalyptic and mystical tractates, Aramaic translations of the Bible, synagogue poetry, magic bowls and amulets.[19] These phenomena do not give the impression of conforming to rabbinic sources and ideas, but this leaves room for the question to what degree the people were attracted to the sages, and to what degree they regarded them as representing a different *Weltanschauung*, with which they had little in common.[20]

I is my belief that the people were mostly convinced that the rabbis were leading them in the right direction. The talmudic greeting 'peace be with you, my masters', and the custom of welcoming a newly arrived sage with food, drink and a bed shows that they were widely regarded as holy men and religious authorities. We do not know how profoundly so-called non-Jewish practices of the people were affected by the rabbis; we do know that they were modified, challenged and changed by them. One could say that the rabbis held much higher standards than the people, but there can be little question that a majority of the community decided which opinions and instructions to accept. Every aspect of Jewish life was a matter of debate, even the combat against idol worship and questions concerning the validity and legitimacy of astrology. In order to exclude or eliminate any possible imitation of the foreign cults they imposed many special restrictions on the Jews, but they did not ban the study of Greek language and science; on the contrary, as a talmudic saying indicates: 'Why use Aramaic in the land of

[18] Babylonian Talmud *Hullin*, fol. 127a.
[19] Sperber, *Magic and Folklore*, pp. 71-118.
[20] Chernick, '"Turn it and turn it again"'.

Israel? [Use] either the language of the sanctuary [Hebrew] or Greek!'[21] They were acquainted with the cults and rites of idolatry, but they assailed the idols mentioned in the Bible and did not denounce the contemporary Greek or Persian gods. It is surely time to accept the recognition that the character of Judaism in the Graeco-Roman world was influenced to a greater or lesser extent by Greek culture to the point that it is not realistic to speak of a Jewish phenomenon as 'essentially Jewish' as opposed to 'hellenistic' or 'hellenised'.[22] Many scholars still have to fight the as-sumption that Judaism clad in Greek dress is not really Jewish, and there will be more to fight against, when new evidence points to Greek being the language of synagogue liturgy in areas like southern Italy, Egypt, Asia Minor, and even in Palestine. The Greek Bible translation of Aquila was commonly used in Late Antique synagogues, therefore Jews may at large have tried to accommodate themselves to the larger Hellenistic world of which they were a part and at the same time reasserted the character of their own heritage within it. When it comes to an evaluation of the Talmud of Babylonia, one often finds scholars who treat the circles of rabbis who produced this work and who are cited in its pages as a self-contained society with little interest in or contact with the world around them.[23] Others have started to take the Talmud's references to pagan religious institutions seriously. However, these studies clearly demand interdisciplinarity on the assumption that there is a continuing debate and engagement with other systems of thought that have left traces even in the absence of explicit citations.

Given the general nature of this contribution I should like to draw an equally general conclusion: Yes, the talmudic sages were interested in ethics rather than metaphysics. They had their rhetorical ways and exegetical means to reach a level of consensus about laws and regulations for the Jewish way of life and for the preservation of Jewish memory and identity.

---

[21] Babylonian Talmud *Baba Batra*, fol. 82a; Schwartz, 'Language, Power and Identity'; De Lange, 'The Revival'.

[22] Lieberman, *Hellenism in Jewish Palestine*; Momigliano, *Alien Wisdom*; Rutgers, *The Hidden Heritage*, pp. 94-95: 'The rabbis ... had categorically declared that the Greek language was one thing, but Greek wisdom another. That this ruling was more or less observed, at least by the rabbis themselves, follows from the fact that Greek philosophical concepts did not find much acceptance in rabbinic literature, that the majority of Greek and to a lesser degree Latin loan words that permeated this literature are to be found in aggadic rather than in halakhic contexts, and that many other loan words concern artifacts used in daily life as well as synagogal furnishings and architecture. In addition, community officials often bear Greek names'.

[23] See also the discussion of the interaction between non-Jews and Jews during the Hellenistic and Roman periods in: Feldman, *Jew and Gentile*; Rutgers, 'Attitudes to Judaism'; Feldman, 'Reflections'.

These masters of the Divine Word served the tradition of the Bible and the Talmud in its own right, but they were influenced by their environment in a thousand greater or lesser ways. The extent of their influence upon actual community discipline was limited, and the will of the community was decisive. As a class, the sages enjoyed recognition within Jewish society as spiritual personalities, yet they had no official position within the community. The significance of rabbinic literature in medieval Jewish existence is a completely different story. The authority of the medieval rabbi has put an indelible stamp upon traditional Jewish life until the early modern and modern period. The opportunities of the last two or three centuries created a bourgeois Jewish culture whose authority derived not from talmudic law but from an ideology of emancipation with lasting consequences for religious authority in modern Judaism. The distinctiveness of the talmudic society cannot be defined by these later standards, but has to be established by synchronic explanation. Only then can one expect a more accurate picture of the pervasive Gentile influence on the 'Jewishness of Judaism' to be drawn.

BETWEEN TERTULLIAN AND VINCENTIUS LIRINENSIS

ON THE CONCEPT *CONSTANTIA VERITATIS* AND OTHER 'CHRISTIAN' INFLUENCES ON SOLINUS

Zweder von Martels

Solinus's *Collectanea rerum memorabilium* is often regarded mistakenly as a work of miraculous tales (*mirabilia*). What the author intended may be gleaned from the following report of Monday, November 12, 2001, on the internet Health Page of the BBC News:

> An Italian woman is due to give birth in a hospital in Rome this week to a baby girl – before returning three months later to have triplets. If both deliveries are successful, it is thought that this will be the first such case in history. Flavia D'Angelo, 20, denies suggestions that she might have had hormone or fertility treatment. She has already named her first child Denise.

The same report gave detailed information on this exceptional phenomenon, and on the risks involved; similar cases, it was said, had never been recorded except in animals.[1] The intense media coverage that followed illustrates the genuine public interest in human stories such as the one taking place in Ascoli Piceno. Flavia D'Angelo's case seemed on the edge of what might just be possible, and yet, unfortunately, it later turned out to be untrue.[2] Nevertheless, it touches on the very nature of men and women. This is why it is 'memorable' – as expressed in the (first) title of Solinus's

---

[1] Sir Thomas Browne (*Pseudodoxia*, pp. 162-166), however, mentions the existence of cases of superfetation in women in Chapter XVII ('Of Hares'). As to animals, he refers to Aristotle (*De generat. animal.*, iv.5.773a), Herodotus (iii.108.3) and Pliny (*Nat. hist.*, viii, 29).

[2] After the Health page of the BBC News internet site had paid attention to the Italian case on 12 November 2001 and again on 18 December, it returned to the subject on 11 March 2002 with the heading: 'Double pregnancy was a sham' and a long further explanation: 'A young Italian couple who astonished the medical world late last year by claiming to have conceived triplets when the woman was already six months pregnant have admitted that the whole story was fabricated … Flavia had not at any time been pregnant. Questions are now being asked as to how such a sham could ever have reached international proportions ...'.

work: *Collection of Memorable Things* (*Collectanea rerum memorabilium*).[3]

The positivistic approach of modern science leaves little space for the irrational. This partly explains why Solinus's former fame has diminished. Elsewhere, I have discussed the reception of Solinus's book prior to Theodore Mommsen's edition of 1864.[4] The great German scholar showed little mercy towards Solinus's intellect and style of writing. Solinus has gradually become an outcast, and he is now almost completely neglected in modern studies on Late Antiquity. It is clear, however, that his work played an interesting role from the second part of the fourth century.

In this paper, I make an attempt to determine Solinus's role in the discussions of contemporary pagans and Christians. The gap between the two sides has long been regarded by many as unbridgeable,[5] but a growing number of studies are opening our eyes to the interactions between both groups. Few cultures are homogeneous, and this certainly holds for the melting pot of Late-Antique Rome,[6] – the place regarded by Solinus as the 'head of the world', and probably the one where he worked and lived.[7]

Vincentius Lirinensis and Solinus

It is useful first to recall an important debate, which, since the fifth century, has played within the Christian church, since it has led me to the new interpretation of Solinus presented below. Cardinal Newman in the nineteenth century and

---

[3] The title page of Arthur Golding's translation of 1587 provides an example of what to expect (antiquated letters in this and the following quotations are modernised): *The Excellent and Pleasant Worke of Collectanea rerum memorabilium of Gaius Julius Solinus Polyhistor. Contayning the noble actions of humaine creatures, the secretes & prouidence of nature, the description of Countries, the maners of the people: with many mervailous things and strange antiquities, seruing for the benefitt and recreation of all sorts of persons.*

[4] See my 'Between Pliny'.

[5] Norden, *Die antike Kunstprosa*, p. 452ff.: 'Die principiellen Gegensätze zwischen hellenischer und christlicher Literatur'; Leeman, *Hieronymus' droom*. In Late Antiquity, too, attempts were made to blur points of agreement between pagan and Christian culture. Arnobius sneers at pagans who tried to protect their traditional, polytheistic, Roman religion against Christian ideas by the suppression of writings of Cicero, who tended towards a monotheistic view – probably his *De natura deorum*. For this, see Cicero, *De goden*, p. 14 (introd.). See also Arnobius, *Adversus nationes*, pp. 115-116.

[6] Hopkins, *A World Full of Gods*; Momigliano, *The Conflict*; Ando, 'Pagan Apologetics'; Edwards, Goodman and Price, *Apologetics*; Lee, *Pagans*; DePalma Digester, *The Making*.

[7] Schmidt, 'Solins Polyhistor', p. 34.

Archbishop Cranmer during the Reformation were among those who took sides in these discussions.[8] The issue at stake was the important question of which Christian group could rightly claim to possess the true, orthodox faith. The answer was found in the interpretation of the three criteria which the abbot Vincentius Lirinensis [de Lérins] had formulated in 438 in his treatise *For the Antiquity and Universality of the Catholic Faith against the Profane Novelties of all Heretics* (*Tractatus pro catholicae fidei antiquitate et universitate adversus profanas omnium haereticorum novitates*). Confronting the heretical views which threatened the unity of the Roman Catholic Church in his age, Vincentius stated among other things:

> ... within the Catholic Church itself we are greatly to consider, that we hold that, which hath been believed, *every where*, *always*, and *of all men*: for that is truly and properly *Catholic* (as the very force and nature of the word doth declare, which comprehendeth all things in general after an universal manner). And that shall we do if we follow *Universality*, *Antiquity*, *Consent*. Universality shall we follow thus, if we profess that one faith to be true, which the whole Church throughout the world acknowledgeth and confesseth. Antiquity shall we follow, if we part not any whit from those senses which it is plain that our holy elders and Fathers generally held. Consent shall we likewise follow, if in this very antiquity itself we hold the definitions and opinions of all, or at any rate almost all, the Priests and Doctors together.[9]

In short, universality, antiquity and consent (*universitas*, *antiquitas* and *consensio*) are, in Vincentius's view, the three interrelated criteria for determining who is holding the true faith. The same three criteria are found in Augustine's writings,[10] but they are also in Solinus, who is the main subject of

---

[8] Lirinensis, *For the Antiquity*, pp. 133-168 (on Cranmer and other sixteenth and seventeenth-century examples); Ward, *The Life*, vol. 2, p. 377.

[9] Lirinensis, *For the Antiquity*, p. 8 (underlining in the following quotations is mine, Z.v.M.): *In ipsa item Catholica Ecclesia magnopere curandum est, ut id teneamus, quod ubique, quod semper, quod ab omnibus creditum est. (Hoc est etenim vere proprieque Catholicum, quod ipsa vis nominis ratioque declarat, quae omnia fere universaliter comprehendit) sed hoc ita demum fiet: si sequamur Universitatem, Antiquitatem, Consensionem. Sequemur autem Universitatem hoc modo, si hanc unam fidem veram esse fateamur, quam tota per orbem terrarum confitetur Ecclesia: Antiquitatem vero ita, si ab his sensibus nullatenus recedamus, quos sanctos majores ac Patres nostros celebrasse manifestum est: Consensionem quoque itidem; si in ipsa vetustate, omnium vel certe pene omnium Sacerdotum pariter et Magistrorum definitiones sententiasque sectemur.*

[10] Cf. Augustine, *De utilitate credendi*, 14.31: *cur non igitur apud eos potissimum diligentissime requiram, quid Christus praeceperit, quorum auctoritate commotus aliquid utile praecepisse iam credidi? tune mihi melius expositurus es, quid ille*

this paper. I recount, first, some general information concerning Solinus; subsequently, the first dedicatory letter, the source of my arguments, is discussed. The final part of the paper is devoted to examples of Solinus's reaction to contemporary discussions.

Solinus and the history and contents of the Collectanea rerum memorabilium *or* Polyhistor

Almost from the beginning, Solinus's famous book circulated in two versions. The first one, entitled *Collectanea rerum memorabilium,* was used by Mommsen for his edition of 1864.[11] The second one was a slightly improved version, corrected by the author. Mommsen mistakenly rejected this corrected version, since he regarded Solinus's second introductory letter (dedicated to a certain Adventus) as the work of an interpolator of a later date. The second letter, however, is now regarded as authentic.[12] In it, Solinus explains that he has changed the original title to *Polyhistor*, the title used in the important editions before that of Mommsen.[13] Peter Lebrecht Schmidt has shown that it is likely that the author lived in the second half of the fourth century, and not in the third as has generally been believed since Mommsen. The main arguments concern Solinus's vocabulary and language, and the fact that, according to a number of manuscripts, Constantius (not Adventus) was the dedicatee of the first dedicatory letter. Schmidt proposed the visit of Constantius II to Rome in 357 AD as a plausible date.[14] Soon after that date, Ammianus Marcellinus, Augustine, and Martianus Capella made use of Solinus's work.[15]

---

dixerit, quem fuisse aut esse non putarem, si abs te mihi hoc commendaretur esse credendum? hoc ergo credidi, ut dixi, famae celebritate, consensione, uetustate roboratae (transl. *The Advantage of Believing*, p. 432: 'Why, then, should I not most preferably seek among them, with all diligence, for Christ's teachings, since it was under the influence of their authority that I already believed that Christ had taught something profitable? Will it be from you that I am to receive a better explanation of what He said, when I would have no belief in His existence, past or present, if it had been you who had commended this doctrine to my belief? In this matter, as I said, I believed a report which had the strength of numbers, agreement, and antiquity'). I am grateful to Mr. Ted Kuiper for drawing this passage to my attention.

[11] See note 3.

[12] Schmidt, 'Solins Polyhistor', pp. 27-35.

[13] No inventory of the more than eighty editions of Solinus exist. Milham ('C. Julius Solinus', p. 77) mentions C. Salmasius and G. F. Muratori as *the* editors before Mommsen; both used the title *Polyhistor*. Interestingly, the title of Golding's English translation (1587) was *Polyhistor*, but G. Kish, the editor of the facsimile edition, changed it in accordance with Mommsen's views; compare note 3.

[14] Schmidt, 'Solins Polyhistor', p. 33.

[15] For these three authors, see Mommsen in Solinus, *Collectanea*, p. 243.

The *Polyhistor*, as we shall call the work henceforth, contains more than bits and pieces chosen at random from Pliny, Pomponius Mela and a few others. The author followed a pre-meditated scheme.[16] The first part treats of the (mythical) history of Rome, called the 'head of the world' (*caput mundi*), in the proud words used long ago by Livy.[17] The foundation of Rome and its further history until the Augustan era are remembered. Augustus's reign is praised as a period free of war and rich in talents. Significantly, the opening lines of the first part provide an etymological explanation – highly respected in Late Antiquity[18] – of the name of Rome, regarded by Solinus as a Greek translation of *Valentia*, as it was named by its first settlers. Solinus pays attention to Caesar's Roman calendar, and to mankind itself in its manifold appearances, before his description of the countless natural phenomena begins. Now and then, thematic digressions on, for instance, dolphins, crocodiles, and horses interrupt the geographical order, which starts with Italy.

The first dedicatory letter: a new interpretation

In the first dedicatory letter Solinus explained his educational programme. Despite this, the letter has been regarded as a piece of little interest. On the basis of Tore Janson's study of *Latin Prose Prefaces*, Schmidt only lists the 'traditional' topoi:

> Von den beiden Briefen ist der chronologisch erste … in der Tat ein eigentlicher Widmungsbrief mit den zugehörigen Topoi, a) der Captatio benevolentiae, b) der Hervorhebung des Texttyps (Kompendium), einer auf Information, nicht Rhetorik zielenden Sprachform, und c) einer Abhängigkeit von bestimmten Quellen, d) der Inhaltsangabe (Geographie einschliesslich der Bewohner, Tiere und seltenen Steine), und schliesslich e) einer Entschuldigung und Rechtfertigung der Beigabe einer kleinen Dosis paradoxographischer Elemente, deren Glaubwürdigkeit – wie die des Gesamtwerkes – bei der Autorität der erlesenen (*exquisita uolumina,*

---

16 'As concerninge Man I have saide sufficient. Now to the intent we may returne to our determined purpose, our stile is to be directed to the recital of places: and chiefelie and principally to Italy' (Golding, transl., *The Excellent*, cap. VII = Solinus, *Collectanea*, p. 31, ll. 6-8). For similar interventions, see Solinus, *Collectanea*, p. 44, l.13: *Flectendus hinc stilus est …*; p. 47, l. 17: *Si respiciamus ad ordinem temporum vel locorum …*; p. 67, ll. 3-4: *Nunc in Thraciam locus pergere et ad validissimas Europae gentes vela obvertere …;* p. 206, l. 3: *Tempus ad Oceani oras reverti …*

17 Livius, i.16: … *caelestes ita uelle ut mea Roma caput orbis terrarum sit*; cf. *idem*, 2.26.

18 Isidore of Seville, renowned for his etymological explanations, was, in fact, building on a long scholarly tradition, which we also find in juridical works.

remoti und doch *receptissimi* auctores) Quellenautoren (*antiquitatis diligentia, vestigia monetae veteris*) liege.[19]

Unfortunately, this quotation demonstrates a form of systematism that prevents readers from recognising the man behind the *Polyhistor*, and discourages rather than encourages research into unexplained features. The letter is less traditional than is thought, and this, I hope, will become clear from the contents of the first twelve paragraphs. After a short *captatio benevolentiae* (1), the author claims (2) that the book is prepared as an abridgement, restrained with moderation so that the multitude of subjects discussed is not extravagant, and the beauty of its style is not unfortunate. (3) 'Reading it closely one will discover that the book, instead of the gold-leafes of eloquence, contains yeast which will make our judgement grow'. (4) 'For this book, I studied several carefully selected books with the intention of keeping my feet away from the more common subjects, and staying with the more remote ones'. (5) 'The larger part of the book is concerned with the record of geographical places; most of its subject-matter is more fit for that part'. (6) 'It seemed proper to mention them so that we should produce the famous places of the earth and tracts of the sea, each in their order, observing the division of the world'. (7) 'We have inserted also a great many things that agree differently so that, if nothing else, variety, at least, remedies the reader's sense of distaste'. (8) 'Among these things we have portrayed the natures of men and other living creatures'. (9) 'Added are a few things on exotic trees, on the form of peoples living in the most remote places, on the different customs of hidden races, and also on several other things worthy to be mentioned; it seemed to be careless to pass over these things, the authority of which – and this I should especially like to recommend to your assiduity – flows from the best received authors'. (10) 'For what could be ours, since the diligence of the ancients has left nothing to stay untouched until our age?' (11) 'Therefore, I ask you, do not judge the trustworthiness of this edition on the basis of the present time, for, following the traces of the ancient stamp, we have preferred to select universal opinions, rather than to alter them'. (12) 'So, if somehow you understand things otherwise than I wish, please forgive my want of eloquence: the steadfastness of truth is found in the authors followed'.[20]

---

[19] Schmidt, 'Solins Polyhistor', p. 27.

[20] Solinus (ed. Mommsen), pp. 1-2 (the paragraph numbering and the underlining are mine, Z.v.M.): 1. *Solinus Advento salutem. Cum et aurium clementia et optimarum artium studiis praestare te ceteris sentiam idque oppido expertus de benevolentia tua nihil temere praeceperim, e re putavi examen opusculi istius tibi potissimum dare, cuius vel industria promptius suffragium vel benignitas veniam spondebat faciliorem.* 2. *liber est ad compendium praeparatus, quantumque ratio passa est ita moderate repressus, ut nec prodiga sit in eo copia nec damnosa concinnitas.* 3. *cui si animum propius*

The author of this letter wrote with an educational end in mind. Some of the manuscripts think of Solinus as a *grammaticus*,[21] that is, as a teacher of grammar, which includes the study of literature and history as the basis of erudition. The expression *fermentum cognitionis* finds its origin in Solinus's pedagogic method, which becomes clear if we look at Pliny the Elder, his main source of information. Pliny, in his *Natural History*, had followed the method of nature. He chose to study life in accordance with its nature: life (or nature) is to a great extent directed by anomaly, and it must be explored accordingly. In general, our knowledge and understanding depend on a great variety of irregular pieces of information. These pieces are like the little stones of a large mosaic, or the dots of a pointillist painting. Each individual stone or dot is without large significance, but receives its function when the human brain is able to connect it with others all comprising a larger image. It is the function of variety to stimulate the human mind to this activity.[22] Pliny – and this also holds for Solinus – seeks

intenderis, velut fermentum cognitionis magis ei inesse quam bratteas eloquentiae deprehendes. 4. *exquisitis enim aliquot voluminibus studuisse me impendio fateor, ut et a notioribus referrem pedem et remotis largius inmorarer.* 5. *locorum commemoratio plurimum tenet, in quam partem ferme inclinatior est universa materies.* 6. *quorum meminisse ita visum est, ut inclitos terrarum situs et insignes tractus maris, servata orbis distinctione, suo quaeque ordine redderemus.* 7. *inseruimus et pleraque differenter congruentia, ut si nihil aliud, saltem varietas ipsa legentium fastidio mederetur.* 8. *inter haec hominum et aliorum animalium naturas expressimus.* 9. *addita pauca de arboribus exoticis, de extimarum gentium formis, de ritu dissono abditarum nationum, nonnulla etiam digna memoratu, quae praetermittere incuriosum videbatur quorumque auctoritas, quod cum primis industriae tuae insinuatum velim, de scriptoribus manat receptissimis.* 10. *quid enim proprium nostrum esse possit, cum nihil omiserit antiquitatis diligentia, quod intactum ad hoc usque aevi permaneret?* 11. *quapropter quaeso, ne de praesenti tempore editionis huius fidem libres, quoniam quidem vestigia monetae veteris persecuti opiniones universas eligere maluimus potius quam innovare.* 12. *ita si qua ex istis secus quam opto in animum tuum venerint, des velim infantiae meae veniam: constantia veritatis penes eos est quos secuti sumus.* 13. *sicut ergo qui corporum formas aemulantur, postpositis quae reliqua sunt, ante omnia effigiant modum capitis, nec prius lineas destinant in membra alia, quam ab ipsa ut ita dixerim figurarum arce auspicium faciant inchoandi, nos quoque a capite orbis, id est ab urbe Roma principium capessemus, quamvis nihil super ea doctissimi auctores reliquerint, quod in novum praeconium possit suscitari, ac supervacuum paene sit relegere tramitem decursum tot annalibus.* 14. *ne tamen prorsus dissimulata sit, originem eius quanta valemus persequemur fide.*

[21] Schmidt, 'Solins Polyhistor', p. 34

[22] The awareness of and predilection for variety are also found in Pliny the Elder, *Nat.hist., Praefatio* 12-13; Gellius, *Noctes Atticae, Praefatio* 3 and 16; Aelianus *De natura animalium, Epilogus*.

this in a mannered style and vocabulary, but especially in the rich contents of his works.

The second aspect of Solinus's letter, which so far has escaped notice, is his connection of criteria which were later used by Vincentius: *antiquitas, universitas and consensio*. The first of these terms appears in Solinus's claim that he has followed the footsteps of the ancients. By being vague about his sources he has gained freedom as an author. What this means will be shown in the last section, below. Nevertheless, he expresses his sympathy for the ancients in different ways. He does this in paragraph 10 with *antiquitatis diligentia*, in 9 with *auctoritas ... de scriptoribus ... receptissimis*, in 11 with *opiniones universas eligere maluimus potius quam innovare*, and finally indirectly in 12 with *constantia veritatis penes eos est quos secuti sumus*, for the authors imitated were the ancient ones mentioned above.[23]

Close examination of this passage in the dedicatory letter shows that not only the words *vestigia monetae veteris*, the *antiquitas* idea, but also the words *opinones universae*, the *universitas* idea, are, just as also happens in Vincentius's text, closely interconnected by a multitude of ways. The recovery of 'universal opinions' is to be considered as the result of wide reading in books carefully selected (4), and in the books of the best received authors (9). The same holds for the uncommon expression *constantia veritatis*, the *consensio* idea, which suggests that, though truth appears in many different forms, it is nevertheless steadfast. With these words Solinus echoes another non-traditional expression, *differenter congruentia* (7). The idea of steadfast truth also returns in *auctoritas de scriptoribus receptissimis* (9), in *opinones universae maluimus eligere quam innovare* (11), and finally, in the remark that *constantia veritatis* is found in the ancient authors followed. What all these closely intertwined phrases show

---

[23] The relation between 'antiquity' and 'truth' (*veritas*) returns several times in Solinus, for instance: Solinus, *Collectanea*, p. 21, ll. 13-16: *nunc si de ipsis hominum formis requiramus, liquido manifestabitur nihil de se antiquitatem mendaciter praedicasse, sed corruptam degeneri successione subolem nostri temporis per nascentum detrimenta decus veteris pulchritudinis perdidisse* (Golding, transl., *The Excellent*, cap. IV: 'Nowe if wee shall mooue question concerning the personages of menne, it will manifestly appeare, that antiquitie hath vaunted no lyes at all of it selfe: but that the ofspring of our time beeing corrupted by succession growing out of kinde, hath through the decrease of them that are nowe borne, lost the comlinessse of the ancient beautie'). Solinus, *Collectanea*, p. 22, ll. 5-6: *scripta quoque ex antiquitate memorias accersunt in fidem veri bello Cretico* ... (Golding, transl., *The Excellent*, cap. IV: 'Also there are wrytings registred in remembraunce of thinges doone in auncient time, which avouch the assuredness of the trueth, wherein it is specified, that in the Candian warre ...').

is, in the first place, the careful and coherent structure of the dedicatory letter. Solinus had put out much work into it. He entered the debate concerning the question of which are the reliable instruments with which to separate truth and untruth. This is not merely a philosophical and theological problem, but one that was also relevant to Solinus. His book contains many subjects seemingly at odds with probability or fact. Incredibility, however, is not what Solinus wanted, as I have explained at the beginning of this article.

This is not the proper place for an extended discussion of the three criteria used by Solinus, but a few lines do need to be drawn between him and earlier and later examples. It is good to remember that the search for truth can lead to two solutions. The first one is absolute verity, for which faith is essential; the second one is a relative truth, based on general acceptance and consensus. Men often waver between the two extremes without finding rest. A few lines of John Donne's third satire 'On Religion' (between 1590 and 1597) suffice to remind us that this has long been a theme in literature:

> Gracchus loves all as one, and thinks that so
> As women do in divers countries go
> In divers habits, yet are still one kind
> So doth, so is Religion; and this blind-
> ness too much light breeds; but umoved, thou
> Of force must one, and forc'd, but one allow,
> And the right; ask thy father which is she,
> Let him ask his; though truth and falsehood be
> Near twins, yet truth a little elder is;
> Be busy to seek her; believe me this,
> He's not of none, nor worst, that seeks the best.
> Satire III.65-75[24]

As to the origin of the expression *constantia veritatis*, it is worth mentioning that Cicero, in his *De natura deorum*, exalted the 'truth' (*veritas*), 'order' (*ordo*), 'system' (*ratio*) and 'steadfastness' (*constantia*) of the heavenly spheres in comparison with the lower regions of the earth.[25] On earth, truth

---

[24] Donne, *Selected Poetry*, p. xxviii (introd.).

[25] Cicero, *De natura deorum*, ii. 56: *Nulla igitur in caelo nec fortuna nec temeritas nec erratio nec vanitas inest contraque omnis ordo veritas ratio constantia; quaeque his vacant ementita et falsa plenaque erroris, ea circum terras infra lunam (quae omnium ultima est) in terrisque versantur. Caelestium ergo admirabilem ordinem incredibilemque constantiam, ex qua conservatio et salus omnium omnis oritur, qui vacare mente putat is ipse mentis expers habendus est. Haud ergo, ut opinor,*

is less stable and appears to be more dependent on consensus and convention. It is, as he stated in his *De officiis*, important for the maintenance of society that 'the foundation of justice is faithfulness, which consists in being constantly firm to your word, and a conscientious performance of all compacts and bargains'.[26] Elsewhere Cicero used the ingenious iuxtaposition *inconstantia veritatis ignoratione* indicating that 'the largest unsteadfastness' goes hand in hand with 'ignorance of truth'.[27]

Tertullian first combined *constantia* and *veritas* in one expression in a discussion on the priority of truth over error in religious matters.[28]

---

erravero, si a principe investigandae veritatis huius disputationis principium duxero. Zeno igitur ... (transl. Rackham: 'In the heavens therefore there is nothing of chance or hazard, no error, no frustration, but absolute order, accuracy, calculation and regularity. Whatever lacks these qualities, whatever is false and spurious and full of error, belongs to the region between the earth and the moon (the last of all the heavenly bodies), and to the surface of the earth. Anyone therefore who thinks that the marvellous order and incredible regularity of the heavenly bodies, which is the sole source of preservation and safety for all things, is not rational, himself cannot be deemed a rational being. I therefore believe that I shall not be wrong if in discussing this subject I take my first principle from the prince of seekers after truth, Zeno himself. Now Zeno ...').

[26] Cicero, *De officiis*, i.7.23: *Fundamentum autem est iustitiae fides, id est dictorum conventorumque constantia et veritas*. (transl. Miller: 'The foundation of justice, moreover, is good faith – that is, truth and fidelity to promises and agreements').

[27] Cicero, *De natura deorum*, i.43: *Cum poetarum autem errore coniungere licet portenta magorum Aegyptiorumque in eodem genere dementiam, tum etiam vulgi opiniones, quae in maxima inconstantia veritatis ignoratione versantur* (transl. H. Rackham: 'With the errors of the poets may be classed the monstrous doctrines of the magi and the insane mythology of Egypt, and also the popular beliefs, which are a mere mass of inconsistencies sprung from ignorance'). Compare also Seneca, *Ad Lucilium epistulae morales*, 102.13: *Putas tu posse unam omnium esse sententiam? non est unius una sententia. Illi placet verum, veritatis una vis, una facies est: apud hos falsa sunt, quibus adsentiuntur. Numquam autem falsis constantia est: variantur et dissident* (transl. Gummere: 'And can you suppose that all men are able to hold one opinion? Even an individual does not hold to a single opinion. With the good man it is truth that causes belief, and truth has but one function, and one likeness; while among the second class of which I spoke, the ideas with which they agree are unsound. Moreover, those who are false are never steadfast: they are irregular and discordant').

[28] Tertullian, *Ad nationes* i.20 (ed. Schneider): *Redigit conscientia veritatem et ad constantiam veritatis* (transl. *The Writings of Tertullian*, vol. I, p. 461: 'Our conscience has returned to the truth, and to the consistency of truth'); for the relation between *constantia*, *veritas* and *conscientia*, see Schneider on p. 318, referring to Tertullian, *Praescriptiones*, 31.4: *Ea sententia manebit adversus posteriores quasdem haereses quibus nulla constantia de conscientia competit ad defendendam sibi veritatem*; cf. Stirnimann, *Die Praescriptio*, pp. 69-70; Michaelides, *Foi*, p. 20ff. (transl. *The Writings of Tertullian*, vol. II, p. 87: 'This sentence will keep its ground

Augustine used it again in his *De consensu evangelistarum* where it, as in Solinus, can be translated with 'steadfastness of truth'.[29] Both Solinus's and Augustine's *constantia veritatis* verge on Vincentius's *consensio*, or 'consent', which was 'unanimity of truth' – not a democratic one, but one on the basis of the two other criteria: *antiquitas* and *universitas*.

In discussions on truth, 'antiquity' became an important argument as well. Cicero, for instance, looked at the examples of the past with deep respect.[30] The inclination to look backwards became conspicuously strong

---

in opposition to all later heresies, which have no consistent quality of kindred knowledge inherent in them – to claim the truth as on their side').

[29] Augustine, *De consensu*, iii.7.30: *si enim hodie quisquam uolens alicuius uerba indicare dicat nomen alterius, a quo dicta sint, qui tamen sit amicissimus et familiaritate coniunctissimus illius, cuius uerba dicere uoluit, et continuo recordatus alium pro alio se dixisse ita se colligat adque corrigat, ut tamen dicat bene dixi, quid aliud intuens nisi tantam inter ambos esse concordiam, inter illum scilicet cuius uerba dicere uoluit et alium cuius ei nomen pro illius nomine occurrit, ut tale sit hoc istum dixisse, quale si ille dixisset, quanto magis hoc de prophetis sanctis intellegendum et maxime commendandum fuit, ut omnium libros tamquam unius unum librum acciperemus, in quo nulla rerum discrepantia crederetur, sicut nulla inueniretur, et in quo maior esset constantia ueritatis, quam si omnia illa unus homo quamlibet doctissimus loqueretur?* (transl.: 'For if today anyone, wanting to quote the words of someone, should mention the name of some other person by whom the words may have been said, who is for all that the greatest friend and most familiar associate of that former one whose words he wanted to quote, and if he, thereupon recalling that he had mentioned someone else instead of the other, should recover his senses and correct himself, with the result that nonetheless he might say 'I have spoken aright', hinting at nothing else that that there is such great agreement between both men, namely, between him whose words he wanted to quote and the other, whose name occurred to him instead of the name of that former one, such that what this one has said is just such as if that other had said it himself - how much the more, in the matter of the message of the holy prophets, is it understandable and to the greatest degree commendable, that we should accept the books of all [prophets] as if the single book of only one, in which no discrepancy of content would be believed to exist still less be found, and in which there would be a greater steadfastness of truth than if one man, however learned, had said all those things?'). After Augustine the expression *constantia veritatis* was used by Hrabanus Maurus, Hildegard of Bingen, and Thomas à Kempis. For Augustine's search for truth in general, see his *Confessiones*, xii.19.28, 23.32, 25.34, 31.42 and O'Donnell's commentary on these passages.

[30] E.g., Cicero, *De oratore*, ii.9.36: *Historia vero testis temporum, lux veritatis, vita memoriae, magistra vitae, nuntia vetustatis* ... (transl. Sutton: 'And as History, which bears witness to the passing of the ages, sheds light upon reality, gives life to recollection and guidance to human existence, and brings tidings of ancient days ...'); Caesar, *De bello civili*, i.6.7: *contra omnia vetustatis exempla* ('contrary to all the examples of ancient times').

after the decline of the Roman empire had set in.[31] Solinus's use of fairly uncommon expressions like '*receptissimis*'[32] and '*vestigia monetae veteris*'[33] express the same idea. However, it goes without saying that the same *antiquitas* idea was not popular with everyone: Tertullian, for example, rejected it in his debate with traditional and conservative pagans.[34]

A third, elusive, idea with which Solinus was playing is the *universitas* idea. Plato had been sensitive to the idea of an invisible, universal, bond connecting the things below and above, [35] and among the Romans Cicero, among others, had continued this tradition.[36] It later played an important role under the Christian authors of Late Antiquity, and this may have inspired Solinus in his use of the expression *opiniones universas*.[37] Though each of the three concepts – *antiquitas*, *universitas* and *consensus* – had a long and complex history, Solinus, apparently, was the first – or was at least one of the first – to connect the three to determine the 'ancient universal truth'.

'Christian' influences on Solinus

Schmidt was the first to cast doubt upon Mommsen's firm denial of Christian influences on Solinus. He gave as one example Solinus's moral rejection of the human sacrifices in Gaul as detestable (*detestabilis sacrorum ritus*).[38] Tertullian's expression *constantia veritatis* might be seen

---

[31] E.g., Aulus Gellius's (*Noctes Atticae*, xix.8.15) coinage of the word *classicus* in the sense of 'superior', or 'of the first rank'; *Digesta Iustiniani*, p. xxxx*: *Constitutio Tanta*, § 10: *Tanta autem nobis antiquitati habita est reverentia, ut nomina prudentium taciturnitati tradere nullo patiamur modo* (transl.: 'Now we had such great reverence for antiquity that by no means did we suffer these men to consign to oblivion the names of those learned in the law').

[32] Cf. Lewis and Short, *A Latin*, s.v. *recipio*: the word *receptus* in the sense of 'received' was post-classical and very rare; only Solinus and Tertullian, *Adv. Marc.* iv.5 (*auctoritas receptior*) are mentioned.

[33] I have been unable to find a similar expression in any other author. But its meaning is similar to the expression *vestigia veterum* popular since Walter of Châtillon (d. 1184?) first used it in his sixth satirical poem (*Cur sequi vestigia veterum refutem*). For the importance of the word *vetustas* in the humanist era, see Witt, *In the Footsteps*, pp. 28, 38.

[34] Cf. Tertullian, *Ad nationes*, i.10.3-7.

[35] See the entry 'Universals' in: *Encyclopedia of Philosophy*, vol. VIII, p. 194ff..

[36] Cicero, *De Oratore*, iii. 20-21; *De legibus*, ii.3.8; i.16.43; ii.3.81, etc.

[37] The expression was once used by Cicero, *De Legibus* i.45.

[38] Schmidt, 'Solins Polyhistor', p. 31, referring to Gaulish human sacrifices (Solinus, *Collectanea*, p. 99, ll. 3- 5), and the addition in a number of manuscripts of the third class *detestabilis sacrorum ritus* (*idem*, p. 218, l. 15ff.). *Idem*, p. 99, ll. 3-5: *Infamantur ritu incolarum, qui ut aiunt (veri enim periculum non ad me recipio) ini-*

as another piece of evidence to the contrary. Exploring this question further, I have found a respectable number of indications that Solinus had undergone Christian influences. On two other occasions, at least, Solinus condemns cannibalism as detestable or criminal (*execrandi / nefarius*), whereas his sources gave no such comment.[39] The savages of Ireland are also treated with little sympathy: they smear their faces with blood of humans whom they have killed.[40] That a different perception of the value of human life and an

---

uria religionis humanis litant hostiis. (Golding, transl., *The Excellent*, cap. XXXIII: 'It is ill spoken of for the custome of this inhabiters, who (as is reported) for I auouch not my selfe to have had triall of the truth, after a detestable manner, (not to the honor, but rather to the iniurie of Religion,) offer men in sacrifice'). Mommsen refers to Pliny, *Nat. hist.*, xxx. 13: *Nec satis aestimari potest quantum Romanis debeatur, qui sustulere monstra, in quibus hominem occidere religiosissimum erat, mandi vero etiam saluberrimum* (transl. Jones: 'It is beyond calculation how great is the debt owed to the Romans, who swept away the monstrous rites, in which to kill a man was the highest religious duty and for him to be eaten a passport to health'); but this seems uncertain as there is little resemblance in the wording, though Pliny indirectly condemns this cruel tradition. Caesar, too, merely mentions this cruel habit as a matter of fact (*De bello Gallico*, vi.16).

39 Solinus, *Collectanea*, p. 82, ll.16-9: *post Anthropophagi, quibus execrandi cibi sunt humana viscera: quem morem impiae gentis adiacentium terrarum prodit tristissima solitudo, quas ob nefarium ritum finitimae nationes metu profugae reliquerunt*. (Golding, transl., *The Excellent*, cap. XXIIII: 'After them are the Anthropophags [explanation added in the margin: 'Meneaters or Cannibals'], who like cursed captives feede on mans fleshe. The which custome of that wicked nation, the Coantryes adioyning beare witnesse of, by lying continuallie waste, the inhabiters of them abandoning them, and running away for feare of that cruell outrage'). This strong moral and religious rejection is absent in Pliny, *Nat. Hist.*, vi.53; Mela ii.1.13. Compare also Mommsen's synoptic survey of these passages (Solinus, *Collectanea*, p. xx); *idem*, p. 84, ll. 15-6: *Inter Anthropophagos in Asiatica parte numerantur Essedones qui et ipsi nefandis funestantur inter se cibis* (Golding, transl., *The Excellent*, cap. XXIIII: 'Among the Anthropophags in the part of Asia are numbred the Essedons, who likewise are embrewed with the same ungracious foode'). Mela ii.1.9 is without any moral condemnation. Solinus, *Collectanea*, p. 131, ll. 8-9, no moral qualification as in Pliny, *Nat.Hist.*, vi.195.
Solinus, *Collectanea*, p. 182, l. 5: ... *Anthropophagi gens asperrima*, no further comments as in Pliny, *Nat.Hist.*, vi.53.

40 Solinus speaks of the culture of the inhabitants of Ireland as ... *inhumano incolarum ritu aspero* ... (*Collectanea*, p. 100, l. 6; Golding, transl., *The Excellent*, cap. XXXII: 'unciuill for the sauage manners of the inhabiters'; I do not see any resemblance to Mela iii.6.6 referred to by Mommsen); Solinus is drawing from an unknown source, where he adds *sanguine interemptorum hausto prius victores vultus suos oblinunt. fas ac nefas eodem loco ducunt* (Solinus, *Collectanea*, p. 100, ll. 9-10; Golding, transl., *The Excellent*, cap. XXXIIII: 'When they haue ouercome theyr enemies, they first besmeere their faces in the blood of them that be slayne, and then drinke of it. Be it right or be it wrong, all is one to them').

aversion from the needless spilling of human blood was growing in Solinus's age,[41] is also clear from another passage, based on Pomponius Mela. The latter recounts that people in India often kill their parents before they die of disease or old age, that they eat them, and that they regard this as an act of piety. He adds that the most prudent ones and those who had studied wisdom, do not wait until death takes them away, but they throw themselves into a fire in advance, 'and glad and glorious they call upon death to appear'. Remarkably, Solinus neither mentions this selfish act, nor Mela's comment. Yet, in reaction to Mela's words on the killing and eating of parents, he adds that they do not regard this *as a crime* but as an act of piety.[42] These words suggest that he and his contemporaries were inclined to regard this as a crime. The addition of these words and Solinus's omission of Mela's positive interpretation of the immoral suicide of some of this people may be explained by new views hostile to euthanasia, and the glorification thereof.

There are other examples of Solinus's sympathy for views popular among Christians. In this vein, he castigates the lack of chastity of the Aethiopians,

---

[41] Another interesting example is found in Eutropius's abbreviated history of Rome, written during the reign of Valens (364-378). Discussing Constantine and the Christian emperors before or after him, the author did not mention Christianity at all, with the single exception of Iulianus Apostata. He tells that this emperor had pursued Christians vehemently, but that (like Marcus Aurelius) he had abstained from blood (Eutropius, x.16).

[42] Mela, iii.7: *Quidam proximi parentes, priusquam annis aut aegritudine in maciem eant, velut hostias, caedunt; caesorumque visceribus epulari fas et maxime pium est. At, ubi senectus aut morbus incessit, procul a caeteris abeunt, mortemque in solitudine, nihil anxii, exspectant. Prudentiores, et quibus ars studiumque sapientiae contigit, non exspectant eam, sed, ingerendo semet ignibus, laeti et cum gloria arcessunt.* (transl. Mela, *Description*, p. 119: 'Some kill their parents (when they are on the verge of decline) like sacrificial animals before the parents decline from age and illness, and it is both morally right and absolutely pious to feast on the viscera of the slain parents. By contrast, when disease or old age have set in, the old and infirm withdraw far from the others and without any fear at all await death in isolation. More prudent individuals, those who are involved emotionally in the practice and pursuit of wisdom, do not wait for death but happily and gloriously bring it on by hurling themselves onto fires'). Solinus, *Collectanea*, p. 187, ll. 3-8: *sunt qui proximos parentesque priusquam annis aut aegritudine in maciem eant, velut hostias caedunt, deinde peremptorum viscera epulas habent: quod ibi non sceleris, sed pietatis loco numerant. sunt etiam qui, cum incubuere morbi, procul a ceteris in secreta abeunt nihil anxie mortem expectantes* (Golding, transl., *The Excellent*, cap. LXIII: 'There are that make as it were a sacrifice of theyr parents and kinsfolke, before they become bare with sicknes or age, and then make a feast wyth their flesh, which thing in that countrey is not counted a wyckednesse, but a godlinesse. There are also that in extremitie of sicknesse, or when diseases lynger uppon them, get themselves into some secrete corner farre from resorte, and there quietlie abyde for death').

calling them degenerate.[43] Such a strong condemnation of their incestuous life[44] is absent in Pliny v. 45, the main source of this passage. Even more interesting are Solinus's remarks bearing on Judaea. They mainly derive from Pliny, as M. Stern concluded, but they also include details from unknown sources. One example is Solinus's mentioning of Gomorrah, unknown in pagan literature before his time; another is the destruction of Jericho by Artaxerxes, which scholars usually connect with Eusebius's *Chronica* (vol. ii.p. 112, ed. Schoene) and authors such as Hieronymus and Orosius who were drawing upon Eusebius.[45] A third example is the passage on the Essenes. Pliny had also spoken about this chaste people, which used to reward innocent men with admission to their country. This policy was so attractive, that this people grew vigorously – even though women were not allowed in.[46] What clearly distinguishes Solinus from Pliny is his praise, as

---

[43] Solinus, *Collectanea*, p. 130, ll. 11-17: *Garamantici Aethiopes matrimonia privatim nesciunt, sed omnibus in venerem vulgo licet. inde est quod filios matres tantum recognoscunt: nam paterni nominis nulla reverentia est. quis enim verum patrem noverit in hac luxuria incesti lascivientis? eapropter Garamantici Aethiopes inter omnes populos degeneres habentur: nec immerito, quia adflicta disciplina castitatis successionis notitiam ritu improbo perdiderunt* (Golding, transl., *The Excellent*, cap. XLII: 'The Garamants of Aethyop know no seuerall marriage, but use their women in common, who that list. Thereon it cometh that the children acknowledge onely their mothers, for the name of father hath no reuerence at all among them. For who is able to knowe hys father, where such incestous lecherie runneth at large. Therfore are the Garamants of Aethyop counted a bastard people among all nations: and not without cause, considering how they haue infringed the discipline of chastitie, and by a wicked custome destroyed the knowledge of their succession').

[44] On ascetism and chastity in the fourth and fifth century, see: Leyser, *Authority*; Brown, *The Body*, pp. 37-39 (on the Essenes); Elm, *'Virgins of God'*.

[45] Stern, *Greek and Latin*, vol. 2, pp. 416-422, esp. pp. 421-422. As yet, it is unclear whether Solinus knew Greek; on the other hand, Hieronymus seems too late for Solinus. As to Gomorrah, it should be added that Stern, under Mommsen's influence, placed Solinus in the third century. In the fourth century Gomorrah remained unmentioned by pagan authors, though Christian ones from Tertullian onwards started to refer to the place.

[46] Solinus, *Collectanea*, p. 155, l. 14 – p. 156, l. 1: *interiora Iudaeae occidentem quae contuentur Esseni tenent, qui praediti memorabili disciplina recesserunt a ritu gentium universarum, maiestatis ut reor providentia ad hunc morem destinati. nulla ibi femina: venere se penitus abdicaverunt. Pecuniam nesciunt. palmis victitant. nemo ibi nascitur nec tamen deficit hominum multitudo. locus ipse addictus pudicitiae est: ad quem plurimi licet undique gentium properent, nullus admittitur, nisi quem castitatis fides et innocentiae meritum prosequatur: nam qui reus est vel levis culpae, quamvis summa ope adipisci ingressum velit, divinitus submovetur. Ita per immensum spatium saeculorum, incredibile dictu, aeterna gens est cessantibus puerperiis ...* (Golding, transl., *The Excellent*, cap. LXVII: 'The inner partes of Iewrye

formulated in the words *maiestatis ut reor providentia*, – 'through the providence, as I deem, of the majesty'.[47] With the word 'divine' (*divina*) added to *providentia*, we have the same expression as the Christian author Firmicus Maternus had used around the same time in his *De errore profanarum religionum*; moreover, we here come close to expressions used by Lactantius.[48] The use of expressions popular in Christian circles is, as we have stated, not exceptional in Solinus.[49]

---

towarde the West, are possessed by the Essaenes: who using a notable trade of dyscypline by themselves, haue departed from the custome of all other nations, ordeyned heereunto (as I thynke) by the prouidence of the diuine maiestie. There are no Women among them: they haue utterly renounced fleshlie lust: they occupy no mony, they live by dates: no manne is there borne , and yet there wanteth no store of men. The place it selfe is deputed to chastitie: whereunto though many resorte from all places about, yet is none admitted unlesse his approoued chastitie and innocent life make him woorthie. For hee that is guiltie of neuer so light a fault, cannot get in there, make he neuer so much sute: but is remooved by the power of God. Thus time without minde (a wonderfull thing to bee spoken) the nation continueth, and yet no chyldbeddes among them'). Cf. Pliny, *Nat. hist.*, v. 73: *Ab occidente litora Esseni fugiunt usque qua nocent, gens sola et in toto orbe praeter ceteras mira, sine ulla femina, omni venere abdicata, sine pecunia, socia palmarum. in diem ex aequo convenarum turba renascitur, large frequentantibus quos vita fessos ad mores eorum fortuna fluctibus agitat. ita per saeculorum milia (incredibile dictu) gens aeterna est, in qua nemo nascitur. tam fecunda illis aliorum vitae paenitentia est!* (transl. Rackham: 'On the west side of the Dead Sea, but out of range of the noxious exhalations of the coast, is the solitary tribe of the Essenes, which is remarkable beyond all the other tribes in the whole world, as it has no women and has renounced all sexual desire, has no money, and has only palm-trees for company. Day by day the throng of refugees is recruited to an equal number by numerous accessions of persons tired of life and driven thither by the waves of fortune to adopt their manners. Thus through thousands of ages (incredible to relate) a race in which no one is born lives on for ever: so prolific for their advantage is other men's weari-ness of life!').

[47] Cf. Parma, *Pronoia und Providentia*; Bos, *Providentia divina*.

[48] Lactantius, *Diuinae institutiones* (*Constantino dedicationes et dualistici loci, secundae editionis*), ii, *additamentum ad cap.* 26.27: *te providentia summae divinitatis*; Lactantius, *De opificio Dei*, 3.2: *prouidentissima illa maiestas*; Firmicus Maternus, *De errore*, 27.5: *Sed hoc ad imaginem futurorum prouidentia diuinae maiestatis inuenit*. It is worth noticing that Golding, either instinctively or on the basis of a Latin text not mentioned in Mommsen's critical edition, translated Solinus's words with *by the prouidence of the diuine maiestie*; cf. note 47, above.

[49] Other examples are words like *mendaciter, tabefacio, inspiratio, mystice, spiritialis, flagranter, flexibilitas, inculpabilis, vagabundus, innubilo, inorno*, etc. For a complete survey, see Hyskell, *A Study*.

Conclusion

Through the examples in this paper, we are able to see Solinus combining interesting and valuable information from ancient sources with new and contemporary ideas so as to stimulate general understanding of the diverse nature of men, animals and other natural phenomena. In the Middle Ages, the *Polyhistor* was read as an illustration of the richness of God's creation. The steadfastness of truth to be found in Solinus gave it something of the character of a universal and eternal Book of Nature. I am inclined to believe that it is no coincidence that this metaphor was introduced shortly after Solinus, first by Chrysostomus, then by Augustine. In any event, the wonderful natural stories of Solinus contributed to the growing popularity of this book in the Latin tradition. It was this idea of the Book of Nature which added new dimensions to Psalm 19, and later became a living reality for humanists like Petrarch and Rudolf Agricola of Groningen.[50]

In conclusion, it has become harder to determine Solinus's position in late ancient society. As with Boethius's *De consolatione philosophiae* the *Polyhistor*, at first sight, raises the impression of being a purely pagan piece of work. But as with Boethius, Solinus may have wished to write a book which is 'the embodiment of human wisdom and knowledge'[51] in the fields described. At least in Boethius's case, we now know that he was a Christian and wrote among other things a small treatise on the Trinity.[52] The picture of Solinus has also changed, now that we have found several words, expressions and ideas betraying an acquaintance with and sympathy for, at least, some 'Christian' standpoints. But perhaps more important is that Solinus's addititions and changes enable us to see that Solinus gave new intellectual and moral dimensions to his sources. This was probably far more appreciated in his own age, the Middle Ages and in the humanist era than we moderns might suppose.[53]

---

[50] For Solinus and the Book of Nature, see my 'Solinus and Augustine'.
[51] On Boethius, see Von Albrecht, *A History*, vol. 1, pp. 1714, 1728-1729.
[52] For the text of this work, see Boethius, *De consolatione.*
[53] I wish to thank Alasdair MacDonald for his observations and comments and for mentioning Vincentius Lirinensis to me some years ago.

THE *CORPUS IURIS CIVILIS*

CODIFICATION AND THE ENCYCLOPEDIA OF LEGAL KNOWLEDGE

Bernard H. Stolte

Introduction

Roman law is one of the original creations of the antique world. Its originality lies in the fact that it is the earliest legal system which was truly a system, in which attention was paid to principles and coherence. It is also the earliest legal system in which some of the participants started to reflect upon the way that system worked. To no other legal system of Antiquity one could justifiable apply terms such as 'legal science' and 'legal scholarship'. In Rome, legal knowledge was considered worth acquiring: not only was it possible to 'learn' law, but there existed a group of people who were 'professionally' engaged in law, who shared the same body of knowledge and whose common activities made it possible to make generally accepted pronouncements about legal questions. In short, Rome knew a legal profession.[1]

In the present context of 'knowledge and learning', the question arises of the way in which legal knowledge and learning were organised. I shall not quibble about either term: the pundits may hold that there is no such thing as legal knowledge in the Popperian sense of the word. The Romans themselves were in no doubt, however: Ulpian, one of their most famous jurists, wrote that 'Iuris prudentia est divinarum atque humanarum rerum notitia, *iusti atque iniusti scientia*'.[2] Pomponius, of whom we possess a mini-history of Roman law, did not hesitate to use the term *iuris (civilis) scientia*.[3] Suffice it to say that we are in excellent company when employing the words *scientia* and *prudentia*, knowledge and learning. We meet them in impressive quantity in Roman legal literature.

---

[1] For a general orientation in Roman legal history see, out of a vast bibliography, e.g., Bretone, *Storia*, to which I refer generally. On the history of ancient Roman legal science – on which term see the next paragraph – one might wish to consult, e.g., Schulz, *History*; Kunkel, *Herkunft*; Frier, *The Rise*. Recent literature about individual jurists of the second and third centuries and their works may be found in the surveys of Roman legal literature by Liebs.

[2] D. 1,1,10,2 (Ulp. 1 reg.).

[3] D. 1,2,2,35 and 40 (Pomp. ench.).

Six centuries of historical enquiry into Roman law, which has its origins in the time of the Renaissance, have yielded a fairly accurate idea of how this knowledge accumulated. For the moment I leave out of consideration the early centuries, during which the law was in the hands of the *pontifices*. Legend has it that the Law of the Twelve Tables of c. 450 BC was the result of a popular movement against the priestly monopoly of knowledge of the law. Using the numerous quotations by later writers it has been possible to produce a more or less satisfactory reconstruction of the greater part of the text of these tables.[4] Through indirect transmission we still have fragments of a commentary on the Twelve Tables by Gaius in the second half of the second century AD. Whatever the shrouds of uncertainty surrounding the Twelve Tables, we may safely say that this early text was being discussed, orally as well as in writing, in order to ascertain the precise meaning of its words for later generations. It is at this early moment that we meet the twin poles of legal discourse: text and interpretation. It is no coincidence that the same legend describes how the people at first felt cheated when they received the Twelve Tables, for the *pontifices* had not disclosed their rules of interpretation, rules which were then 'leaked' to them by one of the secretaries. Knowledge of the law was about the content of rules as well as about the way these rules were applied in practice, for only thus could one predict how they would be applied in future – or at least hope to be able to make such predictions.

It is with these notes on the early history of Roman law that Groningen law students are introduced into the *arcana* of not only Roman law, but of law as such. Since Roman Antiquity, the continental European legal tradition has, in some form or another, depended on the availability of a written set of norms, and has been in the hands of a class of people professionally engaged in the interpretation and application of these norms. Both the substance of the norms and the methods of interpretation are greatly indebted to the jurists of ancient Rome, especially of Rome in the late Republic and the first two and a half centuries of the Empire. It is not in Rome itself and in Antiquity, however, but in Bologna and in the late eleventh century that our modern legal tradition truly begins. The intervening event, essential to the birth of that tradition, is the codification of the law by the Roman or, according to one's preferences, Byzantine, emperor Justinian I in the sixth century AD.

Our theme is encyclopedic texts. The concept of ἐγκύκλιος παιδεία suggests the availability of a body of material purporting to contain all there is to know.[5] Insofar as the Romans acknowledged the concept, they seem to

---

[4] Cf. Riccobono, *Fontes*, vol. I, pp. pp. 21-75.

[5] See, most recently, Christes, 'Enkyklios Paideia'.

have left jurisprudence out of their 'encyclopedias', or at least not to have seen it as a separate entity such as agriculture or medicine. Yet, separate from such collections, they eventually managed to compile an extremely successful encyclopedia of legal knowledge, namely, Justinian's codification. Not only did it claim to contain everything that an educated lawyer needed to know, it also was officially promulgated as an exhaustive collection outside which there was no valid legal knowledge. In other words, according to Justinian's intention no one would be allowed to invoke a rule of law which had not been included in the codification.[6]

The idea of a codification as an exhaustive collection of laws does not come as a surprise. After all, it is quite common for a legislator to claim a monopoly on legislation, or even to have such a monopoly attributed to him in a constitution (*Grundgesetz*). As far as formal legislative activity is concerned, no one besides the legislator would be allowed to exercise it. True, we are used to quoting Montesquieu for the *trias politica*, but the notion of a legislative monopoly is much older. In that sense Justinian's codification is unremarkable.

Much more interesting, and more relevant to our theme than Justinian's view of the emperor as the formal *fons omnis iuris*, is his claim to be the source of all legal knowledge, too. For a correct understanding of this claim we have to take a closer look at the constituent parts of his codification.

In the later Roman empire, until Justinian's day, it was possible to invoke in a court of law rules originating from two different areas. One body of material consisted of the *leges*, the imperial laws or *constitutiones*, which had been issued by successive Roman emperors. The other body was made up of the writings of Roman jurists, whether they were commentaries, casebooks or monographs, and this was known as *ius*. Justinian codified – i.e., brought together within a *codex* – the relevant *constitutiones*, abrogated those which he considered out of date, arranged the remaining ones according to subject and declared the result to be the exhaustive, exclusive *Codex Justinianus*, promulgated in 529. The writings of the Roman jurists were selected and arranged in a comparable way in 533 in the *Digesta* or *Pandectae*, and at the same time even a text-book of Roman law for beginners was issued, the *Institutiones* or *Elementa*. In 534 a second, revised edi-

[6] The best source for the nature of Justinian's codification is the emperor's own words in the preambles to the constituent parts (see below), easily found in the standard edition: *Corpus iuris civilis*, vols i-ii, ed. Mommsen and Krüger, numerous reprints since 1872-1877. For a survey of Justinian's codification, with a certain emphasis on its claim to exhaustiveness, see Van der Wal and Lokin, *Historiae*, pp. 31-38. There is an ocean of literature on the subject.

tion of the *Codex Justinianus* completed the codification; the first edition was thus made obsolete after only five years and it has not been preserved.

The compilation of the *Codex Justinianus* or Code was the logical outcome of imperial legislative activity and was not without precedent: the emperor Theodosius II had issued his *Codex Theodosianus* in 438, which served as a model to Justinian. In retrospect, the promulgation of the *Digesta* or Digest was a much greater event, though hardly noticed as such by Justinian's contemporaries. It is on this part of the Justinianic codificatory enterprise that I concentrate in the remainder of this paper.

The Digest is above all a massive monument of legal learning of a bygone age. To avoid all misunderstanding, this statement represents what was already the truth in Justinian's own time. One only has to consider the facts. Divided into fifty books, each book subdivided into *tituli*, the Digest contains a selection of legal opinions on the subjects mentioned in the rubrics of these titles, each opinion or fragment of an opinion scrupulously attributed to its original author by means of an *inscriptio*. The most recent jurists occurring in the Digest were Hermogenianus and Arcadius Charisius of c. 300, both represented with only a handful of fragments. The latest author represented by a substantial amount of writing was Herennius Modestinus, who lived in the first half of the third century. More than two centuries separated Justinian from the most recent of the authors who, had they not been incorporated in the Digest, would in many instances have been no more than a name. In order to understand what was being codified in the Digest, we have to move back a couple of centuries and take a closer look at these jurists.

The distinctive character of Roman law, as already mentioned, consisted in the fact that it was a veritable system, designed by a class of people who used the same technical language, and who argued within that system after a proper method. These people, the Roman jurists, were consulted on legal problems and gave their opinions about them, an essentially oral activity, but one which naturally lent itself to be documented in writing. During the Republic, the authority of this segment of the aristocratic classes was such that in their discussions Roman law could be seen to develop: intervention by a legislator was comparatively rare. This is not the place to discuss the mechanism in detail, but innovations were often brought about by jurists who managed to adapt existing and widely accepted views to a novel set of facts and formulate a reasoned opinion on the case in hand. On the basis of their authority such an opinion would be accepted by the *praetor*, whose tribunal was the formal channel through which legal 'novelties' found their way into public life. Insofar as they were subject to debate, this debate took place within the closed circle of jurists, and their writings enable us to follow the discussion in a number of cases. Characteristically, in

Rome one became a jurist by becoming a pupil of an established jurist and following him about when he was giving advice and taking part in these debates. Roman legal learning, one could say for this age, resided in the body of legal opinion available to Roman citizens; though oral by origin and nature, it lived on in written form: no doubt individual opinions were noted down, but usually they were transmitted in collections of opinions, of the existence of which we have abundant proof.[7]

Roman law at this time was a typically casuistic system. The sum total of innumerable cases and their solutions, in many cases with alternative solutions, made up the standard against which a new case would be measured. Legislation played a subordinate role. In addition to the casuistic literature which was occasioned directly by the social role played by the jurists, other forms of literature take stock of the state of the law: they describe certain areas in a more systematic form.

If, however, we were to look for an encyclopedic text in the field of law of Republican Rome, we should look in vain. The essence of legal learning was not knowledge of every case that had ever been dealt with, but the ability to reason within the system. The system had been given a certain backbone, first by the Twelve Tables, and later by the praetor's Edict – and Cicero tells us that in his time a schoolboy still had to learn the Twelve Tables by heart – but that had nothing to do with the training of a jurist. A truly casuistic system does not easily lend itself to encyclopedic treatment: the idea of continuous addition and change is incompatible with a static description of 'all there is to know'.

Those living in modern Anglo-American jurisdictions will have read here a description of a familiar situation. There is in fact a close parallel between the law of Republican Rome and the system of the Common law, though with one important difference: in the modern Common law development is in the hands of the judge, whereas in Republican Rome it was carried forward by the jurists who acted in an advisory role. Both systems are characterised by the absence of an over-enthusiastic legislator with codificatory aspirations, aspirations that on the Continent are usually attributed to Napoleon and since his time have never disappeared. Napoleon's ideas, however, were far from original in this respect. *L'empereur des Français* consciously took Justinian as his model. In order to understand the attraction of that model one has to understand the main lines of the development of Roman law during the first five centuries of the Empire.

---

[7] Cf. the so-called *Index Florentinus*, a working-list (?) of the legal literature still available in Justinian's day and explicitly referred to in const. Tanta/Δέδωκεν 20, which has been transmitted with the only complete manuscript of the Digest, now in the Biblioteca Medicea-Laurenziana, the codex Florentinus S.N.

The troubled times of the civil wars of the last century before the era under discussion resulted in a change of more than just the political system: among the social changes they brought about, one sees a diminished role for the old aristocracy, the order to which the jurists traditionally belonged. The jurists, who for reasons of social prestige had been able to exercise an informal but very important influence, gradually entered the bureaucracy surrounding the emperor. The process took a considerable time: it is not without significance that the last great jurists did not die peacefully in their beds, but fell victim to their displeased sovereigns, as the Roman emperors had by then become. The first two centuries AD are the time in which the Roman jurists changed from the status of independent private counsel to becoming the powerful heads of the imperial chancery. In this capacity they still wielded power, but now under firm supervision.

The loss of independence inevitably affected the way in which Roman law developed. Roman private law could still be described as a casuistic system, but with one important difference as compared with that under the Republic. From the third century onwards, answers to legal questions were no longer given by a private individual, but by or at least in the name of, the emperor. The imperial *rescriptum* had become a major vehicle for the formation of the law. When, in the second half of the second century AD, the jurist Gaius enumerated the formal sources of the law,[8] the *leges* (in the sense of the laws adopted by the *comitia* of Republican Rome) and *plebiscita* were no longer active instruments of legislation; the *senatus consulta* were no longer independent of the will of the *princeps*; the *constitutiones principum* were by far the most important source; the *edicta magistratuum* counted only in name; and, together with the *constitutiones principum*, the *responsa prudentium* were in fact the only living sources of law, i.e., springs still flowing. However, as a living source the latter dried up in the course of the third century.

The process of development of Roman law is thus characterised by a gradual monopolisation of the formation of the law by the emperor. In the terms used before, the role of *lex* increased at the expense of *ius*. Yet it is the *ius* that should occupy us here, since that is the repository of Roman legal learning or 'science'. I would suggest that precisely through the intervention of the emperor the *responsa prudentium* were able to become an encyclopedia of legal learning; that that encyclopedia is the Digest; further, that, having been given the form of a codification, that encyclopedic text was prevented from developing in a natural way; and finally, that the availability of an encyclopedic text which at the same time was a 'closed revela-

---

[8] Gai. Inst. I,2.

tion' had a lasting effect on the development of continental legal thought. I conclude by annotating that role as 'closed revelation'.

The emperor and the responsa prudentium

As was said above, Gaius speaks of the *responsa prudentium* as a formal source of law, i.e., one was able to quote an opinion of a Roman jurist in court in the same way as that in which we would cite a paragraph from a code. Obviously this gives rise to the possibility of conflicting opinions being quoted, a problem that had attracted the attention of the emperor. In Gaius's words:[9]

> The answers of the learned are the decisions and opinions of those who are authorized to lay down the law. If the decisions of all of them agree, what they so hold has the force of *lex*, but if they disagree, the judge is at liberty to follow whichever decision he pleases. This is declared by a rescript of the late Hadrian.

The passage raises various questions, but this much is clear: legal opinions expressed by a jurist could have force of law, and at least the emperor Hadrian had paid attention to the phenomenon. We are further informed by the second-century jurist Pomponius, in a casual remark in his treatise on the *successio prudentium*, that the emperor Augustus had been the first to accord a *ius publice respondendi*, an imperial patent to answer legal questions on his majesty's authority.[10] Although many questions remain, apparently from the start the emperors had reined in the free formation of a body of authoritative legal opinion. When, after the first half of the third century, the steady flow of legal writing came to a standstill, the sheer amount of literature from the past remained a problem. In the fourth and fifth centuries the emperors intervened to regulate that mass of legal writing, until finally, in 426, the so-called law of citations of Theodosius and Valentinian[11] selected five 'top lawyers' and promulgated rules of precedence among their opinions in case of conflict.

Implicitly they thus abrogated the authority of a number of 'lesser' names. This might have deprived us of a great deal of historical information about the development and variety of legal opinion especially during the late Republic and early Empire, since the lack of legal authority also means

---

[9] Gai. Inst. I,7 (but cf. Just. Inst. 1,2,8), in the translation by De Zulueta, *The Institutes*, p. 5.

[10] D. 1,2,2,48-50 (Pomp. ench.)

[11] CTh 1,4,3.

the lack of the need to copy and hand down these writings. That the latter have nevertheless been preserved is due to the compilation of the Digest on the order of Justinian I in 530 (*const. Deo auctore*). The passage in which, three years later, on the occasion of the promulgation of the finished result, Justinian says that many of the volumes used came from the private library of his right-hand legal officer Tribonian,[12] is more than just a piece of miscellaneous information: it explains something in the line of transmission of legal writings which would otherwise have been difficult to accept without further comment, namely that, despite their abrogation by the 'law of citations', so many works were still to be found a century later.

The Digest as an encyclopedia

Much has been written on the compilation and organisation of the Digest.[13] I shall deal with one aspect only: that it is also an 'encyclopedia' and a peculiar one at that.

This requires a move in the spirit to sixth-century Constantinople, a cosmopolitan but mainly Greek-speaking city, capital of an empire, the main strength of which lay in the East, where Greek was the *lingua franca.* In such a city and for such an empire Justinian promulgated an anthology of legal writings from the past, almost entirely in Latin. Whatever may be the full explanation of this curious choice of language, part of it surely is the fact that the emperor was consciously reaching back to a glorious Roman past: in his own words, *nosque Aeneadae vocamur.*[14] The ideologically indispensable foundation of his restoration of the empire *armis legibusque*[15] was made visible in the use of the Latin language. This decision was in itself sufficient to make the Digest the summing up *par excellence* of an intellectual effort of twelve centuries, the alpha and omega of legal learning, in short, a sixth-century encyclopedia of *jurisprudentia*, the first and last of its kind. Three observations may be offered by way of explanation.

First, to characterise Justinian's codification as an encyclopedia of legal knowledge is more apposite in the case of the Digest (and Institutes)[16] than in that of the Code. While one may admit that the Code is indispensable – there are actually some cross-references from the Institutes and Digest – one

---

[12] Const. Tanta/Δέδωκεν, 17. Pugsley ('The Florentine Index') explains the *Index Florentinus*, mentioned above, as 'Tribonian's library'

[13] See, e.g., Honoré, *Tribonian*, but cf. also Osler, 'The Compilation'; Pugsley, *Justinian's Digest.*

[14] Nov. 47 pr.

[15] See, *inter alia*, the preamble to the Institutes, *const. Imperatoriam, pr.*.

[16] The Institutes are best taken together with the Digest. They are a text-book for first-year students and preparatory to the Digest.

may safely say that only the Digest could pretend to offer a systematic treatment of the law; the Code, by contrast, only contains the results of imperial intervention. Since this, as has been explained above, had never been systematic, but had depended on occasion and expedience, serious lacunae were inevitable. The various imperial constitutions, to be sure, were offered to the reader in groups according to subject-matter – the *tituli*, the same system[17] also underlying the Digest – but the arrangement according to chronology instead of a coherent pattern of thought applied to the constitutions within a title of the Code made it impossible to consider it other than what in fact it is: useful, indeed necessary information, but by its nature fragmentary and incomplete in itself.

Second, the curriculum which Justinian prescribed[18] for the law schools of Rome and Beirut heavily emphasised the Digest: the students were to spend about three and a half years out of five in studying the Digest under the guidance of a teacher, an *antecessor*. The extensive remnants of this teaching have been preserved for the greater part in the scholia on the *Basilica*.[19]

Third, the choice of the Latin language for the codification must have greatly contributed to the idea that this form of legal writing belonged to the distant past, that it was a 'closed revelation' and that no new knowledge could be added. To which we may add that Justinian himself in various ways formally ruled that there was nothing to add unless by himself: the *Novellae post Codicem constitutiones*.[20]

An encyclopedia that was prevented from developing in a natural way

In his act of promulgation of the Digest Justinian inserted an explicit ruling that only this collection of *ius* could be quoted as an authority in a court of law, on the same level as an imperial constitution, and further, that it should be regarded as the emperor's own words.[21] To put it differently: in the Digest the emperor speaks to his subjects in a long constitution, with the words of lawyers from the past. No opinion or treatise that had not been included in the Digest was admissible in court.

---

[17] Cf. Krüger, *Geschichte*, pp. 377-379.

[18] Const. *Omnem*, §§ 2-6; see also below.

[19] Scheltema, Holwerda and Van der Wal, *Basilicorum libri LX*; Van der Wal and Lokin, *Historiae*, chapt. iii and pp. 90-91.

[20] Being *constitutiones* the Novels were not only in a format different from that of the established genres of Roman jurisprudence, but soon they were also in Greek. The Novels are usually quoted after the edition by Schöll and Kroll in *Corpus iuris civilis* iii.

[21] Const. Tanta/Δέδωκεν,10; 19; 20a; *const. Deo auctore* 7.

We should note that, as a consequence, the Digest could only be changed or added to with the emperor's co-operation. In other words, a revised edition could only come about through a legislative act. Roman legal learning was thus frozen in 533 AD.

Moreover, Justinian imposed a so-called ban on commentaries.[22] At first sight, this measure seems to emphasise more strongly the end of Roman legal learning, but difficulties remain. A normative legal text cries for commentaries and will in fact continuously be commented upon until its replacement or other loss of normative status. Attempts have been made to interpret this curious measure in various ways. The most original and ingenious explanation was proposed by Scheltema, who concentrated on the rest of the paragraph and suggested that the emperor simply prohibited the writing of commentaries on the text in the same manuscript and in the same letters as the main text, which would thereby give rise to confusion between text and commentary; all other forms of commentary would remain permitted without restriction.[23]

Two comments may be made on his suggestion. First, if a literal ban on commentaries was meant, it was singularly ineffective. Lots of commentaries have survived, although very few in the margin of Latin manuscripts.[24] Contemporaries may therefore have seen anything but a literal ban in the passage. Second, however, the idea of a literal ban has proved greatly appealing to later legislators; after all, there is a certain attraction in going down in history as the legislator who made all other legislation superfluous, and a commentary, by its very nature, suggests an imperfection of clarity which would stand in the way of such a claim. As late as the eighteenth century, Frederick II of Prussia issued an actual literal ban, and Montesquieu's conception of the judge as *la bouche de la loi* was after all but a different conclusion from a similar line of reasoning in which the idea of a perfect law is essential.

Whatever precisely Justinian's wishes have been – and one need not hesitate to attribute the wildest ambitions to the emperor – the result was the Digest: a systematic voluminous text, with formal claims to completeness, and promulgated at a moment and in a format which in themselves would already suggest that completeness. Let us now examine the consequences.

---

[22] *commentarios illi [sc. nostrae consummationi*, i.e. *Digestis seu Pandectis] applicare', const. Deo auct.* 12; cf. Tanta/Δέδωκεν 21.

[23] Scheltema, 'Das Kommentarverbot Justinians'.

[24] Two papyri have come to light: P. Reinach 2173 and P. Heid. Lat. 4.

The encyclopedia as codification: the consequences

As has already been noted, the texts incorporated in the Digest were the logical outcome of the nature of the development of Roman law during the Republic and the early years of the Empire. That path of development had come to an end when social and political circumstances had ceased to allow an independent legal profession to be the vehicle of change in legal thought. The idea of collecting and selecting those texts and of presenting them as one systematic whole, thereby incorporating everything one needed, indeed was allowed to know about the law as an intellectual discipline both suitable for teaching and citable in court as a source of law, was a master-stroke attempted, but never successfully completed before.

There is a fair degree of consensus on the transition of Antiquity into the Middle Ages if we are not too particular about dates. As far as the law is concerned, a good case could be made for *Justinianus bifrons* as both an end and a beginning.[25] The emperor's retrospective collection of legal writing and imperial constitutions summed up the legal learning of Antiquity. His mode of promulgation made it clear that only through an imperial act of law could change take place. Venerable *Antiquitas* was now definitely over. Justinian, we may remind ourselves, lived conveniently close to the magic years of 476 and 500 to permit us to cast him in that role. But if the end of Antiquity is no problem, the beginning of the Middle Ages is less self-evident, and here we have to distinguish.

In Byzantium Justinian's reign marks a profound change, both by producing the codification and by showing, in the way it was taught and applied, that the literary Latin culture which it presupposed was insufficiently available. The availability of knowledge of Latin in sixth-century Byzantium has been seriously underestimated,[26] but it was never prominent enough to sustain a Latin legal culture which would answer to Justinian's ambitions. The alternatives produced by teaching and legal practice, in the form of a host of translations, commentaries and summaries in Greek, actually became the basis of what we know as Byzantine law.[27] Crudely phrased, it was Roman law in Greek shape, and in that shape, with many ups and downs, it even survived the fall of Constantinople in 1453. One

---

[25] See also my 'Justinianus Bifrons'.

[26] The Justinianic codification in itself provides a formidable body of evidence, with Tribonian as a towering figure. Two examples from a different sphere are the grammarian Priscian and the poet Corippus, who, though born elsewhere, at some point made a career in Constantinople.

[27] For details of the transformation, see Van der Wal and Lokin, *Historiae*, chapt. iii.

thing, however, is certain: never in any moment of its history did Byzantine law manage to surpass the intellectual qualities of its great Roman ancestor, the 'classical' Roman law of Antiquity with such qualities visible even in the Greek versions of the Digest. The encyclopedia was never spiritually digested, if the wordplay may be permitted.

It is probably also true to say, however, that the Byzantines never replaced Justinian's legislation at the level of the ideal encyclopedic text. They substituted Greek versions, as already said, and reorganised those versions in about 900 in the *Basilica*,[28] but there was never a comparable text in the centre of their legal culture.

One of the more striking differences between Byzantium and the western Middle Ages is found in the fact that in Byzantium it was precisely the Digest – which in my view is the true encyclopedia of Roman legal learning – that was the least successful part of the Justinianic heritage. This brings us to the reassuringly familiar, the 'western' part of the story, to which the remainder of this paper is devoted. The awakening of legal studies in Bologna has been linked with the so-called Renaissance of the Twelfth Century.[29] The story of the 're-emergence' or 'rediscovery' of the Digest has been told many times. It still leaves us with some perplexities,[30] but the main point to remember here is surely the fact that, after some five centuries of almost complete silence, the Digest resurfaced and, in the second half of the eleventh century, became the heart of a new legal curriculum, which gave rise to the University of Bologna.

It is possible to describe the role of the Digest in the genesis of our modern continental legal tradition without going too much into legal detail. Medieval lawyers found in the Digest a vast number of cases and very few clear rules. Almost every imaginable situation seemed to have been dealt with, but frequently in the form of a discussion of possible solutions, solutions rarely appealing to fixed authorities other than the jurists who had advocated them. The result was a collection incredibly complex to handle. It was more sophisticated than any legal text known to them, but also one of which the authority must have been dubious: what was the basis of the validity of the Justinianic legislation in northern Italy in the late eleventh century? There is little use in reviewing the answers to that question which have been thought up in retrospect. To be sure, the notion that Roman law possessed a certain traditional authority had never been lost in Italy. Had

[28] *Ibidem*, chapt. vii.

[29] Kuttner, 'The Revival', describes the emergence of the new jurisprudence in this context. For a different perspective see Berman, *Law and Revolution*.

[30] The certainties and perplexities have been summarised recently in an excellent review-article by Jacobs, 'Die große Zeit'.

that notion not been present, it would have been difficult to concede any basis for its acceptance without serious discussion. Still, it is my guess that at first the Justinianic legislation, and in particular the Digest, was considered as nothing more than very useful teaching-material. The reasons for that usefulness are not difficult to see.

Here was an all-encompassing book, a legal ἐγκύκλιος παιδεία. The medieval manuscripts of the Digest usually contain only one of its three preambles, the *constitutio Omnem*, in which Justinian offered the Digest to the professors of the Byzantine law schools and set out how it was to be taught to the students. It explains how students should start with the *Institutiones* by way of an introduction, but then for more than three years were to concentrate on thirty-six out of the fifty books of the Digest, until in their fifth and final year they would read the Code. I am not for one moment saying that the medieval Bolognese teachers simply followed Justinian's instructions, but the availability of such an all-encompassing book, which could fulfil the same role as the Bible in the theological curriculum, must have been convenient.

Whether or not Justinian had compiled the Digest as an encyclopedia of legal learning, the medieval lawyers had no difficulty in recognising it as such, and what is more, they treated it in just that way. Justinian had emphasised that here was an end to the reigning confusion: this was all there was to know, the definitive summing-up of *iuris prudentia*, and, des-pite its enormous size, one without contradictions,[31] a boast that the med-ieval lawyers took literally and one which led them to explain the book as a whole by the use of the scholastic methods.

It is interesting to observe that the process of expounding and commenting on the Digest in its relation with the Code, which resulted in the mid-thirteenth century in the *glossa ordinaria* of Accursius, had been preempted by the Byzantines by half a millennium, with one important difference: in Byzantium it had been an end, after which new and less ambitious aims were pursued, while in Bologna it became the beginning of the modern legal tradition, spreading over Latin-speaking Europe. It now remains to give an explanation of how the Digest could retain its role of encyclopedic text.

The 'closed revelation': a text impervious to time or change

As we have seen, Justinian had closed off the path of development of 'classical' Roman law through successions of advising jurists, who could add their opinions to those of their predecessors. Of course, he did not prohibit

[31] Tanta/Δέδωκεν,15.

legal advising; it was simply a matter of changing the status of such legal advice. The question whether one would be able to quote a private legal advice as an authority in court depends on various factors. Certainly it does not help if that possibility is more or less ruled out by the explicit claim of the emperor to monopoly status as a source of law. One expression of that claim had been the compilation and promulgation of the Digest as a codification.

As was said above, it is interesting, but here perhaps not very fruitful, to ponder the basis and the extent of the binding force of the Justinianic legislation and especially of the Digest during the High Middle Ages and afterwards. On the other hand, the curriculum of the law faculties of Bologna and elsewhere until at least the French Revolution shows this continual heavy emphasis on the Digest, which in the present context is proof of the vitality of the Digest's role as an encyclopedic text of legal learning. Indeed, that role stands out all the more prominently if we remind ourselves that it was sustained in the face of a changing legal world: inevitably the world changed over the centuries and it would be absurd to imagine that Roman law continued to reign unchanged. It would be correct to say, however, that the Digest managed to maintain its unrivalled position of encyclopedic introduction to legal learning. It is precisely for that reason that the continental legal tradition presents a fairly uniform face until this very day.

The history of this apparently timeless encyclopedia can only be sketched in the broadest lines. These broad lines allow us to see precisely how the timeless aspects came to be emphasised over the centuries.

In the Middle Ages the question of the already considerable age of the Digest hardly played any part at all. Once the Justinianic legislation had been established in the legal curriculum, it was taught with a healthy disregard for the historical dimension. Rather than lose themselves in historical problems, the professors explained the text as if Justinian was still alive. Those portions of the text dealing with problems that were simply no longer relevant to medieval circumstances were omitted, but otherwise texts were explained as much as possible in such a way as to make them applicable to actual problems. One can see why the Digest (and Institutes) lent itself better for that approach than the Code and the Novels. The last three books of the Code, for example, so obviously presuppose the state and administrative system of Late Antiquity that it would be hard to make them relevant for, say, the city council of Milan in the fourteenth century, and this is the reason why initially they were not included at all.

Another good example of how the Digest could be used as the repository of legal doctrine, in relation to the imperial legislation as found in Code and Novels, concerns the law of succession. A considerable part of the Digest is devoted to problems of succession, i.e., heritage, legacies etc. In two

Novels, issued in AD 543 and 548 respectively, Justinian himself thoroughly revised the system of succession: broadly speaking, he replaced the ancient Roman system of succession in the male line by our modern system of succession according to degrees of blood relationship. The result was that the extensive discussions of successory problems in the Digest presupposed a system no longer in force after Novels 118 and 127. Still, the doctrinal discussions in the Digest were hardly affected by the novelty of a different order of succession. The method and principles remained the same, although details might have changed. Indeed, if the order of succession was to change again, as it did in some respects under the influence of canon law, the timeless aspect of the underlying doctrine became all the more obvious.

It is obvious that not all problems occurring in the later Middle Ages would have a close similarity to problems met and dealt with in the world of Late Antiquity. Where such similarity was lacking, no passage in the *Corpus iuris civilis* would be of immediate use and reasoning by analogy would be impossible. Here, too, we have to differentiate between the Code and Novels on the one hand, and the Digest on the other. The latter contained at least an apparatus of principles and conceptions with which the case could be described and analysed, so that solutions to these problems would fit into the legal system as a whole.

It is this relation between timeless doctrine and accidental change that lies at the heart of the continued role of the Digest. The so-called *ius commune*, reigning from the later Middle Ages until the modern codifications of the nineteenth century, is an illustration of this phenomenon. Only with this in mind can one understand that it took so long to dislodge the Digest from its central position in legal thought.

The first step towards a different perspective was made by the Renaissance humanists.[32] The central notion here is historical interpretation. The humanists of the fifteenth to the seventeenth centuries were the first to ask themselves what the original context of a Justinianic passage had been, a question unthought of in the Middle Ages. As noted above, this was the beginning of the history of law as an academic discipline, although the humanists themselves had never had that intention. Inevitably, however, historical investigation led to the conclusion that the *Corpus iuris civilis* had been meant for a society which no longer existed in the early modern period. Again the Digest suffered less from this approach than the Code and the Novels. This is not to say that the Digest did not receive the same amount of attention: on the contrary, I should almost say the reverse. The point, however, is that legal doctrine as a system of reasoning – the Digest –

[32] On humanistic jurisprudence see, e.g., Troje, *Graeca leguntur*; Stein, 'Legal Humanism and Legal Science'.

becomes less easily outdated than legislation about actual problems – Code and Novels – once one has become aware even to some extent of a historical dimension.

In the present context I restrict myself to a few words on the tradition in legal historiography to picture the humanists as the first generation of lawyers to make painfully clear the antiquatedness of Roman law. This undoubtedly is correct, but the statement is incomplete. The effort of historical investigation by the humanist lawyers must have compelled them as lawyers intellectually to distinguish between essential and accidental in Roman law as part of the legal system in force. For my part I find it infinitely more satisfactory to imagine that they were, consciously or unconsciously, able to do so, rather than to have to consider them as schizophrenic persons, torn between medieval legal tradition and humanistic invention.

The process of distinguishing between essential and accidental, between timeless and ephemeral, received a further impetus in the school of the natural lawyers of the seventeenth and eighteenth centuries.[33] The notion of natural law, a philosophical assumption positing a legal system independent of time and place and connected with the very existence of human nature, naturally – I mean, inevitably – raised the question of its relation to positive, man-made law, and thus to, e.g., Roman law. The easiest answer to that question was the identity of ideal and example, because Roman law could then be considered the perfect example of a system based on natural law. In other words, if one were to sit down and write out a system of rules in accordance with what one would consider natural law, one would automatically end up with the Justinianic legislation. Among the protagonists of natural law, this position was taken by Grotius. To be sure, it would be impossible seriously to maintain that position unless, here too, one were to distinguish between essential and accidental, and thus between the Digest on the one hand and Code and Novels on the other.

Again a few notes only. Grotius's position was increasingly abandoned in the school of natural law. Criticism of Roman law became more outspoken and contributed not a little to the genesis of modern law codes in the wake of the French Revolution. Nevertheless it is possible to trace an extensive influence of Roman law in many notions of the natural lawyers. I would suggest that there are two interlocking reasons for this. One is the 'law of gravity': Roman law was established in jurisprudence and legal practice, and one could only ignore it at the risk of becoming irrelevant. The other is the simple fact that, in order to be really independent of the tradition of Roman law, one would have to devise an entirely new legal terminology. As long as one continued to reason in categories stemming from, or at least

---

[33] See, e.g., Thieme, *Das Naturrecht*.

known to, Roman law, the substance of Roman law would not go away. Once more we have to remind ourselves that the distinction between essential and accidental is evidently relevant.

One of the major contributions of the natural lawyers to legal theory is undoubtedly the way in which they mapped out the law into logical groups and sub-groups in an attempt to deduce its rules from first principles. As far as private law is concerned, this distinguishes their approach from the presentation of Roman law in the Digest. The German Historical School of the nineteenth century is in that respect a continuation of eighteenth-century natural law[34] and in 1900 it eventually produced the *Bürgerliches Gesetzbuch* as a normative system of German private law, described in abstract terms which had been extracted from the writings of the Roman jurists collected in the Digest. The present Dutch Civil Code of 1992 is another good example of a code of law answering, if somewhat belatedly, to the demands of systematisation which had been voiced by the natural lawyers of especially the eighteenth century. The substance of this Civil Code, however, would be immediately recognisable to the Roman jurist, who would be able to take part in a legal discussion in the twenty-first century, because the essentials would be familiar to him. Just as the Roman jurist would not feel left out, present-day continental lawyers, or, perhaps more correctly, lawyers from countries with a civilian tradition, would feel at home despite differences of language, because they all belong to a tradition so heavily indebted to the Digest. Which brings us back to our point of departure, the relation between codification and the encyclopedia of legal knowlege.

Conclusion

The first half of this paper offered an explanation of why a truly encyclopedic text in Roman law is not found before the Justinianic legislation, and why in that legislation the Digest is the encyclopedia of legal knowledge. The second half described the continued role of the Digest as that encyclopedia, a role above all visible in its central place in the legal curriculum. It remains to say a few words in explanation of that phenomenon by way of a conclusion.

An all too simplistic and on the whole erroneous explanation would emphasise the intrinsic qualities and merits of the Digest. On the contrary, its faults are many and various, and make its success all the more incomprehensible to the modern level-headed observer. The Digest is one of the most difficult books of Antiquity, impossible to handle without the intermediary of an *interpres*. The first successful interpreters in

[34] Koschaker, *Europa*, p. 254ff., esp. pp. 275-281.

the medieval West were the Glossators, who explained the Digest (in the context of the entire Justinianic legislation) as an encyclopedia of legal knowledge. One could understand the Digest with the *Glossa ordinaria* as the medieval edition of that encyclopedia: the result of the struggle of the first generations of Bolognese interpreters to come to grips with their material. The subsequent generations – unimaginatively called the Post-Glossators or Commentators – established the relation of that encyclopedia to the real world: they brought it to bear on actual problems in contemporary northern Italy. The Post-Glossators introduced another novelty as well: legal treatises in a format independent of the gloss. One hesitates to call the latter monographs, a literary genre that did not come into its own until the humanists needed a suitable vehicle for their invective. But it is certainly the case that various kinds of commentary, which are found in rudimentary form in the glosses, now grew into full-sized treatises for which no room was to be found in the margins of the manuscripts of the *Corpus iuris*. If such room could have been found, I wonder whether it would have been considered the right place: the lawyers from the late thirteenth and fourteenth centuries may intuitively have felt that they were no longer commenting on Justinian's words, but were building on the foundations laid by them. The physical separation of that work from the *Corpus iuris* with the Gloss was beneficial in two respects: it liberated the medieval lawyers from the shackles of an authoritative text, but at the same time it made it possible to maintain the Justinianic text in its role of encyclopedic text.

I am not for one moment denying that all sorts of other factors have influenced the history of the *Corpus iuris* in legal scholarship and practice. Were one, however, to look for an encyclopedic text in this field, nothing would better answer that description than Justinian's Digest. We have seen that it was given an excellent start by virtue of the fact that it was promulgated as a codification of what, in the sixth century, was considered the ideal summing up of legal learning. The combination of a codification and imperial antecedents has been astonishingly successful.

In the final analysis Justinian has managed to rule from the grave even to this day. Of course, the imperial antecedents have long ceased to be relevant. The nation-states of modern Europe have long ago succeeded to the role of legislator par excellence. What remains is a civilian tradition in continental Europe and a number of countries influenced by that tradition, which still makes it useful to study the roots of that tradition. The need to explain the contents to a succession of generations from a perspective of contemporary relevance has always compelled students of Roman law to concentrate on what best lent itself for that purpose: the Digest as a repository of legal doctrine. From that perspective the Digest is still *the* encyclopedic text of the continental legal tradition.

BETWEEN LITERARY TRADITION AND CULTURAL CHANGE

THE POETIC AND DOCUMENTARY PRODUCTION OF DIOSCORUS OF APHRODITE

Jean-Luc Fournet

Dioscorus of Aphrodite is a well-known figure in the field of papyrological studies, but at the same time he is hardly ever mentioned in histories of Greek literature and culture. Dioscorus became a familiar name among scholars of papyri from the discovery of a jar filled with papyri in the village of Kûm Ishqâw, ancient Aphrodite, Middle Egypt, in 1905. This jar contained classic literary works by Homer and Menander, documentary material and, even more remarkably, occasional poems contemporary with this documentary archive. Papyrologists soon realised that the key figure to this collection was a certain Dioscorus, the writer of part of the documentary texts, the owner and reader of the books, and the author of the occasional poems. The study of this archive provides us with the opportunity to reconstruct the family of Dioscorus, his career and his background. Born c. 525 into a Coptic family which was so influential in Aphrodite that it practically held a monopoly over the village, Dioscorus conducted his affairs as a landowner as well as those of his village. Therefore, the papyri of this archive shed a most brillant light upon sixth-century Egypt in all its different aspects.[1]

Papyrologists have generally recognised the great value of the documentary texts of Dioscorus's archive as a source for their historical knowledge. However, these scholars are more reserved, and even sceptical, in their evaluation of his poetic production. This poetic œuvre encompasses fifty-one compositions at present,[2] the majority of which are written in dactylic hexameter, while some are in iambic trimeter. With the exception of some poems on mythological subjects, these poetical writings primarily

[1] For further details on Dioscorus and his background, see the excellent studies by Maspero, 'Un dernier poète grec' and by Bell, 'An Egyptian Village'. The most recent synthesis is by MacCoull, *Dioscorus of Aphrodito*.

[2] See Fournet, *Hellénisme* (=*P.Aphrod.Lit.* IV).

belong to the genre of the *enkômion*, or eulogy: the poems are addressed to prominent persons on the occasion of, for instance, their installation in a public function or their birthdays; some of these writings are epithalamia. The poems have only survived in the form of rough copies in the poet's handwriting on sheets of papyrus which had often been re-used. Since their discovery and their partial publication by Jean Maspero, these poems have been subject to the mockery of the academic world. It was not until the 1980s that they began to be judged less negatively and even became regarded as objects of interest. Nevertheless, Dioscorus has remained merely a footnote in histories of literature, and he has most often been ignored by Hellenists not concerned with the study of papyri.

I believe, however, that Dioscorus's papyri may well illustrate the issues of knowledge, learning and cultural change that are central to this volume. In the first place, Dioscorus's work signifies a perfect paradigm of a rural elite culture in a hellenised province of the Late Roman empire; a culture which can be understood and reconstructed in a quasi-archeological way thanks to this papyrological find. Furthermore, the scope and variety of this ensemble make it possible to understand the processes of acquiring knowledge and learning: the books of Dioscorus's library shed light on the development of the pupil or the reader; the drafts of the poems very concretely show how this acquired culture was subsequently employed for the production of new materials; the documentary archive illustrates the use of literary culture for practical and pragmatic ends. Finally, being embedded in particular traditions, his writing testifies to changes that affected the Greek culture in that period. As will be demonstrated, Dioscorus well exemplifies the tension between tradition and change, tension which brings forth a new culture of writing.

Traditions in poetry and rhetoric

An examination of the poetry of Dioscorus reveals first and foremost that his poetic writings fitted in with the traditions marking the Greek culture of Late Antiquity. For one thing, Dioscorus chose to express himself in verse, and particularly in dactylic hexameter. Thus, he was related to, and was the last representative of, a poetic movement current in third- and fourth-century Upper Egypt: what can be called the poetic school of the Thebaid, as brilliantly analysed by Alan Cameron.[3] After all, did Eunapus not state in his *Life of Sophists*, 493, that the Egyptians were absolutely mad about poetry: ἔθνος ἐπὶ ποιητικῇ μὲν σφόδρα μαίνονται? In fact, it was Upper Egypt that provided late Hellenism with its last great poets. I only mention

[3] Cameron, 'Wandering Poets'.

those who were most important or rather whose works have been partly preserved: Triphiodorus (by 300), author of *The Taking of Ilios*; Cyrus of Panopolis (fifth century), writer of epigrams surviving through *The Palatine Anthology*; Pamprepius of Panopolis (fifth century), author of several lost poems and of an *Ekphrasis* preserved in a Vienna papyrus; Colluthus of Lycopolis (fifth/sixth century), writer of *The Rape of Helen;* Christodorus of Coptos (491/518), author of an *Ekphrasis* (*Palatine Anthology II)*; Musaeus, who is named Θηβαῖος in manuscripts (sixth century), author of the delightful *Hero and Leander*; and especially Nonnus of Panopolis (fifth century), to whom we owe a *Paraphrase of the Gospel According to John* in hexameters as well as a lengthy epic poem consisting of forty-eight books, *The Dionysiaca*. Over the past few decades many contributions have been made to a re-evaluation of this epic poetry from Late Antique Egypt and of its strong influence on Greek poetry from the rest of the Empire. There is no need to dwell on it here.

The writings of Dioscorus not only add to the corpus by Theban poets, but they also are intrinsically influenced by some of them. Apart from Homer, the person who had the greatest impact on Dioscorus was Nonnus of Panopolis (fifth century), the greatest representative of the Theban poetic movement. There are numerous quotations of words, word groups and even nearly complete lines from *The Paraphrase of the Gospel According to John* and from *The Dionysiaca*.[4] After Nonnus, the work which had the strongest influence on Dioscorus is (less unexpectedly) the *Paraphrase of the Psalms* by Pseudo-Apollinarius of Laodicea,[5] who is strongly suspected to be an Egyptian.[6] Musaeus and Pamprepius were also read by Dioscorus, since their texts are cited sporadically by him.[7] Thus, Dioscorus appears to have been quite familiar with the Theban poetic tradition, from which he consciously sought inspiration, and which he perpetuated in his own way.

The other tradition that Dioscorus's poetry thrives on is encomiastic rhetoric, the quasi-obligatory mould for all literary production in that age. His work is in accordance with this tradition on every level: occasion, content and ornament.

First of all, the occasion, or, to use the terminology of ancient rhetoricians, the *hypothesis*. Dioscorus follows the prescriptions outlined by treatises of epideictic rhetoric, in that he composes his poems on the occasion of events that traditionally constitute the *impetus* for a public speaker and that are connected with power, i.e., at a provincial level, the *adventus,* mar-

---

[4] Cf. *P.Aphrod. Lit.*, p. 678, note 51.

[5] *Ibidem*, p. 679, note 54.

[6] Cf. Golega, *Der homerische Psalter*, p. 171; Cameron, 'The Empress', p. 239.

[7] Cf. *P.Aphrod.Lit.*, p. 680, note 56.

riage or birthday of the governor. Thus, half of the poems of Dioscorus are *enkômia* of *adventus*,[8] *enkômia* of birthdays,[9] and epithalamia.[10] The other half belongs to another category: the petition-like poems which I shall discuss below.[11] But in fact, some of these pieces, while being petitions, celebrate at the same time the *adventus* of the emperor or governor to whom they are dedicated (*P.Aphrod.Lit.* IV 10; 11), thus showing Dioscorus's intention of utlising the major occasions for encomiastic rhetoric. Dioscorus reproduces the same scheme even for inferior officials taking up their posts (*P.Aphrod.Lit.* IV 16).

The content of the poems is entirely reliant upon the conventions outlined by theorists of rhetoric (Menander Rhetor and Pseudo-Dionysius of Halicarnassus, author of the *Tekhnê peri tôn panêgurikôn*). One finds the same *topoi* in the poetry of Dioscorus, dealt with in a very similar way, although in a rather chaotic order.

Even in his ornamentation of discourse, Dioscorus relies upon the rhetorical conventions of the eulogy: particular enumerations of comparisons (*synkriseis*) or mythical *exempla* used by the poet can be traced back to the handbooks of rhetoric, and reveal that Dioscorus applied these conventions quasi-automatically.[12]

As is indicated by a passage from his writing, Dioscorus had probably read the works of Menander Rhetor.[13] It is possible that he even owned his treatises. In any case, thanks to the information given in a papyrus in Berlin, it is known that the work of this rhetorician circulated throughout Byzantine Egypt.[14] Dioscorus, however, may have known this author only through *compendia*, perhaps studied at school. It needs to be noted that the two traditions of poetry and rhetoric, which I have just mentioned, were part of the system of learning: poetry and rhetoric were core subjects taught at the last two levels of classical schools, and consequently they became the established modes of expression for any author; two modes of expression that became intertwined as in the work of Dioscorus, and that determined the entire cultural production of the period.

---

[8] *P.Aphrod.Lit.* IV 17-20; see also 39 and 40.
[9] *P. Aphrod.Lit.* IV 21-24; 31(?).
[10] *P.Aphrod.Lit.* IV 32-36; 37(?), 50.
[11] pp. 110-111.
[12] See, for example, the annotations to *P.Aphrod.Lit.* IV 5, 11-12; 7, 16-17; 33, 11-20.
[13] See the annotations to *P.Aphrod.Lit.* IV 5-7.
[14] Cf. Maehler, 'Menander Rhetor and Alexander Claudius'.

Transformation and subversion: christianity and self-appropriation

I shall not further elaborate upon these two aspects of the poetry of Dioscorus, for they are generally known. In my view it is more interesting to discuss the transformations or the subversions to which these traditions are subject based upon Dioscorus's poems and which, since they may not be particular to Dioscorus, must be considered as illustrations of a change in sensibility and the cultural modes of the period.

The first transformation has to do with the appearance and development of Christian themes. It is not appropriate to speak of innovation in this respect. Since the fourth century, poetry had dealt with Christian subjects, as is clear from the poems in the *Codex des Visions* of the Bodmer Foundation,[15] or as can be gathered from the fact that Apollinaris of Laodicea and his son translated the Old and New Testament in verse.[16] Moreover, the two metaphrases by Nonnus and Pseudo-Apollinaris challenge the presumption that using Christian themes in Homerising poetry was an anomaly. However, in any case, these works illustrate a pagan setting – or rather a classical setting – of Christian content. The work of Dioscorus gives us the opportunity to become aware of a most remarkable phenomenon: a hybrid form of Christian and classical elements that seem to me characteristic of Protobyzantine art. Thus, in the eulogy for the chancellor of the praetorium prefect (*P.Aphrod.Lit.* IV 7), a quotation from the Psalms, adapted to the requirements of metre, precedes a comparison with Orpheus, Nestor and Achilles:

τοῖον ἔπος κατάλεξον ἕως παρεμύθετο θυμός·
ὅττι τέττιγξ πολύυμνος ἔχει δέμας ὁπλọ́ṭ[ατον],
νυκṭαδίῃ μελεδῶνι Θεὸν κατ' ὄρεσφι λιγαί[ν]ει.
Κḷụτὸς εὐκλείης βασιλεὺς Θεὸς [ὕ]μνον ἀκούει.
...
'Ορφέα Καλλιόπης ἐνίκησας ἐτητụμί[ῃσιν].
Μῦθον ἐπ' ἐννεσίῃσι ṭεὸν ποθέουσ[ιν ἔπαρχοι.
Νέστωρ οὐ λάθεν ἄλλọς, ['Αχιλλε]ụ̀ς ἄμμ[ι]ν ἐτύχθης
ἱστάμενος προβολ [..]. [.......] ἔπλεο τοῖσδε
νόσφιν ἀλιφροσύν[ης

[15] See *Vision de Dorotheos*, papyrus Bodmer XXIX (ed. Hurst, Reverdin and Rudhardt) and *Codex des Visions*, Papyri Bodmer XXX-XXXVII. For further details see Fournet, 'Une éthopée'.
[16] Sozomen, *Ecclesiastical History*, V 18; Socrates, *Ecclesiastical History*, III 16.

(If people told me that I was too audacious in praising you), here is what I would say while my heart encourages me: 'The cicada, despite its prolific sound, has a very tiny body. However, watching by night in the mountains, it sings of God. And God, the illustrious king of fame [cf. Ps. 23. 7-10[17]*] listens to its hymn ...'. Verily, you have surpassed Orpheus, the son of Calliope. The prefects look for advice from your speeches. You could pass for a new Nestor. You are actually our Achilles, raising up as a rampart ... without his vanity ...*

In another epithalamium (*P. Aphrod. Lit.* IV 34), the Christian God appears next to Dionysus and the divinised Nile:

> Ἀκλινέως κατέμαρψαν ἐοικότε Βελλεροφόντῃ.
> Μήνιν ἀμφιέπουσιν ἀγαλλομένην ὑμεναίοις,
> [π]υρο̣φόρου χαρίεντας ἐπ' ἄνθεσιν αὔλακας αἴης,
> ῥηιδίως Διόνυσον ἐνὶ στεφέεσσι κ(αὶ) Νεῖλον̣.
> Ῥυσάμενος πολύτεκνον ἀεὶ Θεὸς ἐγγυαλίξοι
> οὐλομένης ἀπάνευθεν γάμον προφερέστατον ἄλλω(ν).

They had no trouble finding one another, similar to Bellerophon.[18] *They benefit from the favours bestowed by the Moon, who rejoices in their marriage, from the favours bestowed by the furrows of the wheat-bearing earth which adorn themselves with graceful flowers, by the diligent Dionysos with his crowned head and by the Nile. Offering protection against a fatal misfortune, may God grant them an eternal marriage that surpasses all others and from which numerous children will be born.*

A reversal of the phenomenon occurs in Dioscorus's prose petitions, the prosaic form and the elocutionary situation not necessarily implying a recourse to pagan stereotypes – quite the contrary. For instance, in one of these petitions (*P. Cair. Masp.* I 67020, 5-6), Dioscorus refers to Zeus shortly after having evoked God:

> ... μετὰ τὸν δέσ[π]οτ(α) Θεόν, εὐεργέτα καὶ τὸν (...) προστάτα πάσης τῆς ἐπαρχείας, ὡς ἀπὸ Διὸς γένους λελάχατε αἷμα τῷ ἀληθ εῖ.

[17] Pseudo-Apollinarius uses the same biblical expression in his *Paraphrase of the Psalms* XIII 17,22.

[18] For more information on the figure of Bellerophon, originally classical but transformed by Christianity, see Fournet, 'Un nouvel épithalame', pp. 77-78.

(...You are...) after God our master, our benefactor and patron of the whole eparchy, because, truly, you are of a blood which springs from the race of Zeus.

This phenomenon of hybridisation even takes place on a lexical level: the best example is the subversion created by Dioscorus assigning a Christian significance to the Homeric expression οὐ κατὰ κόσμον which normally has an almost contrary meaning. In a poem in which he celebrates the installation of the new governor, the duke of Thebaid (*P.Aphrod.Lit.* IV 26), Dioscorus writes in line 8:

Ἤλυθες οὐ κατὰ κόσμον ἀεὶ σκέψαι ναετῆρας

You have come to protect the inhabitants for eternity οὐ κατὰ κόσμον.

Giving this expression the Homeric meaning ('in a disorganised, indecent manner') would result in absurdity. Given the context, οὐ κατὰ κόσμον can only have positive connotations. Yet, in the Christian era the word κόσμος ('order') could signify 'the world as opposed to God's order'. The duke thus seems to have come to protect the inhabitants according to principles that do not correspond to the order here below.

Even if this hybridisation of motifs belonging to two cultural worlds so far apart may seem remarkable, it is not surprising: paganism is tolerated as long as it is cultural and not religious.

However, a Christian sensibility lies behind this classical form which implicitly questions the role of this 'antiquarian' paganism. This is revealed by the subversion on the part of Dioscorus of one of the essential figures of rhetorical discourse: the comparison. Any comparison with grand figures from classical mythology is thus modified by an emphasis on the inferiority of the classical figures in relation to the person praised in the poem. For example, in *P.Aphrod.Lit.* IV 33 one can read:

Εὐρώπην σὺ̣ φέρεις εἰς οἰκίον, οὐκ ἐπὶ πό[ντῳ̣],
Λήδης λ̣[έκτρα φέρεις, ἀλ]λ̣' ο̣ὐ̣ πτε̣ρ̣ύγεσσι κομ̣[ίζει]ς.
Νυμφίε, μ̣ὴ .τ ......τ̣ε̣σ̣..ονυ. κ̣αὶ γὰρ 'Α̣πόλλω[ν]
ἡδυβόλων πολύυμνος ἐδέξατο κέντρον 'Ερώ̣[των]·
Δάφνην γὰρ δι' ἔρωτας ἀεὶ Δαφναῖος 'Απόλλων̣.
Δάφνην καὶ σὺ φέρεις, ἀλλ' οὐ φεύγουσαν, ἀκοίτην.
Μύρια Φοῖβος ἔτευξε καὶ οὐκ ἐτύχησεν ἐρώτω(ν).

It is a Europa that you lead to your house, and not over the sea; it is a Leda that you marry, without transporting her on wings. O fiancé, do not [...], for the beautifully singing Apollo himself has also felt the sting of sweet-darting Loves: because of his love for Daphne, Apollo is forever 'the Daphnean'. You as well, you take as your wife a Daphne, who will not flee.[19] *The works of Phoebus have been numerous without his knowing the workings of love.*

This modification of the *synkreiseis* could certainly be considered proof of a change in sensibility caused by the triumph of Christianity. Even more illuminating is *P.Aphrod.Lit.* IV 34, 4:

> Κύπριδος οὐκ ἐνόησεν ἀεικέα δ' ἔργα τελέσσαι
>
> *(Isaac and his wife) have never envisaged accomplishing the indecent works of Cypris.*

Breaking with the conventions of rhetoricians (Menander Rhetor, 400, 5; 402; 6; 404, 25; 407, 7; 11; 409, 7; 411, 13), Dioscorus does not represent Aphrodite in the positive role of the divinity of marriage, but as the incarnation of wanton desire. It is in this line that one may best detect the changes in the traditional rhetorical models brought about by Christianity.

Nevertheless, classical Greek culture continued to be the referential framework for men of quality. The new Christian culture was not considered as oppositional, but complementary to pagan culture. One can therefore speak of a bicultural world, or, more precisely, of a Christian culture which integrated the classical heritage, which appropriated classical models and authorised explicit references to ancient paganism, inasmuch as it was not given a religious meaning, but was considered as a sign of culture. This tolerance points to the irreversible triumph of Christianity, and at the same time to a 'patrimonial' respect for the ancient Hellenic culture.

The other process of transformation that is manifested by the work of Dioscorus has to do with the strongly private nature of his poems. Put differently, Dioscorus subverts the encomiastic poetic tradition by appropriating this tradition for personal ends. He has done so in two ways:

1. There is a clear subversion of the tradition of Theban historical *enkômia*. This genre of *enkômia* constituted one of the most characteristic features of the Theban poetic school, as has been discussed above and as has been

[19] The sentence can also be constructed as follows: Δάφνην καὶ σὺ φέρεις, ἀλλ' οὐ φεύγουσαν ἀκοίτην 'You take a Daphne who will not flee her husband'.

enunciated by T. Viljamaa.[20] These *enkômia* are epic poems depicting Thebaid as subject to barbarian attacks (principally by the Blemmyes), but finally as being saved by a general whose achievements in war are celebrated. Dioscorus takes over this conventional scheme, but transfers it to his own personal misfortune. To give some examples:

a) The Thebans greet their saviour. This theme can be found in two historical *enkômia*:
Heitsch Suppl. 10.9:

Θ̣[ήβ]η̣ μ̣ὴ τ̣ρ̣ο̣μέοις, οὐκ ὄρχαμος ἄλλος ἀμ̣[είνων]

Thebes, do not tremble any longer, there is no better leader.

Heitsch XXXVI r°a, 16-17:

ἐνναέται Θήβ̣ης σταχυωδέο[ς] ὑμνείουσιν
[.... τειρ]ο̣μένοισιν ἀλεξίκακόν σε φανέντα

The inhabitants of Thebes' wheatland celebrate you who appear ending the sufferings of those who are hard-pressed by [...]

Compare these passages with *P.Aphrod.Lit.* IV 10, 21-24, which describes the Thebaid greeting the new duke, Athanasius.

Δέχνυσο, πότνια Θήβη̣, [ὃν ο]ὐκ ἐδάμ[ασ]σαν α.[
ν̣[α]ὶ̣ τὸν 'Αθανάσιο̣ν, κ̣ρ̣[α]τ̣ε̣ρὸν ῥυτ̣ῆ̣[ρα] πο̣λ̣[ήων]·
οὐ γὰρ ἔτι Βλεμύων γένος ὄψεαι, οὐ Σαρακηνῶ(ν),
οὐ τρόμον ἀνδροφόνου ληίστορος ὄμμασι λεύσῃς.

Receive, august Thebes, him who has not been tamed by the [...] yes! Athanasius, the most forceful saviour of cities. For you will no longer see the tribes of the Blemmyes nor of the Saracens. You will no longer witness the fear caused by the murderous looter.

However, Dioscorus's intention becomes clear at the end of the same poem (line 45):

χεῖραν ἐμοὶ ἀτάνυσσον ἐμὴν πενίην διαλύειν

---

[20] See Viljamaa, *Studies*, pp. 45-54.

Stretch out your hand to me to end my poverty.

b) Dioscorus alludes to the Blemmyes in a eulogy for a duke of Thebaid (*P.Aphrod.Lit.* IV 11, 81-82):

> Τοὔνε[κα γ]ουνάζομ̣α̣ί̣ [σ]ε πανάλκιμον Ἡρακλῆα,
> πήμ̣[ατ'] ἀ̣ποπτύε̣ιν [Β̣]λ̣εμύων γένο[υ]ς ἤτοι βοηθῶν̣,

That is why I beg you, almighty Heracles, to banish the evil caused by this tribe of the Blemmyes, I mean the adjutores.

The poet compares the *adjutores* ('assistants'), who plunder his fields, with the Blemmyes, who once terrorised the Thebaid. Thus, one can speak of a transference of a theme characteristic of the historical *enkômia* to the present circumstances of the poet.

What takes place is a narrowing down of perspective, from the collective to the individual. The arrival of the general saviour of the Theban people that occurs in the *enkômia* has been transformed into the *adventus* of the duke to whom Dioscorus appeals for help; the call for help from the Thebaid has turned into the request for assistance that Dioscorus makes to the authorities; the plundering Blemmyes have changed into extorting local officials. Dioscorus thus preserves the schematic framework of the Theban *enkômia*, but fills this framework with a content marked by personal interest.

2. This subversion for personal ends goes even further as it affects the genre of Dioscorus's poetry. In spite of belonging to the tradition of encomiastic poetry, his poems are less concerned with praise than with the assertion of self-interest. A major part of these poems are similar to petitions. Take, for example, *P.Aphrod.Lit.* IV 1:

> Ὦ πτολίαρχε μέγιστε βοηθόε πᾶσιν ἀνάγκης,
> κλῦθι πονιομένου Παφίης χθονὸς ἐνναετῆρος,
> δέξ̣εο 'μῆς γενιῆς τὰ δυσίμερα δάκρυα μόχθων·
> πολλά μοι ἐν γραφίδεσσι χαράγματα οἴκοθεν ἤχθη,
> ὅττι καὶ Γαβριήλις χερείο̣να τῶν πρὶν ἔερξεν
> Πενταπολίτης Θεόδωρος ἀτάσθαλα ἔργα καὶ αὐτός
> ἡμετέρων σφετέρισσεν ἀλωῶν καρπὸν ἀπούρας.
> Χῶρον ἅ̣παντα θέριζε μελισταγέων σταφυλάων.
> Θρέμματα ἠδὲ βόας πόρεν 'Αρσᾶ̣ κτήματα πάντα,
> οὕν̣εκεν ἑνδεκάτης Θεοδόσσιος ὧν λάβε χρυσῶν,

[ἡμετ]έρης γενιῆς βιοτήσιον. Νῦν δὲ φαεινῶν
[σοῦ πρ]οκυλ[ι]νδόμενος πόδας ἰχνῶν, ὕψος ἄρειον

O greatest ruler of the city, general assistance against distress, listen to an inhabitant of the land of the Paphian (=Aphrodite*) in his suffering. Receive the sad tears of pain from my family. I have brought from home many documents testifying that Gabrielius has committed worse crimes than the acts of violence formerly accomplished by Theodorus of Pentapolis: he himself stole and appropriated the fruit from our threshing floors; he reaped entirely the honey-dripping grapes of our vineyard. Flocks, cattle, our entire property he gave to Arsas, because of the sums of gold due for the eleventh indiction, which Theodosius had stolen from us, the resources of our family. And at present, knelt at your radiant feet, O martial highness,* ... (the ending has not been written).

I have proposed elsewhere to consider these petition-like poems as pieces which accompanied the petitions in prose delivered to the authorities.[21] If Dioscorus is by far the best representative of petitions in verse, a parallel in Latin, poem XIII by Sidonius Apollinaris, seems to demonstrate that the genre was more widely used than we could have thought on the basis of the extant literary sources and papyri. In any case, this genre shows how the encomiastic poetic tradition may have been taken over and employed to assert the claims and to voice the needs of a citizen. The poetry thus seems to have shifted from its earlier function to a concern with a subject matter traditionally reserved for the document. This phenomenon, as illustrated mainly by the work of Dioscorus, reveals the way in which knowledge, in a poetical-rhetorical form, became invested with a practical significance and in which a culture of erudition, mingled with archaism and artifice, changed into a mode of acting engaged in the present.

The emergence of a new culture of writing

These changes mark the emergence of a new culture. One of its characteristics is the primacy of writing. Encomiastic poetry and rhetoric had been essentially oral mediums of expression and part of a communal system: the *enkômia* were traditionally declaimed in front of a community assembled for important events. Dioscorus shows us that encomiastic poetry could be delivered in a written form and be addressed to a single individual: the compositional structure and function of his poems give us strong reasons to believe that they were not recited in public, but sent or presented as written

[21] Cf. *P.Aphrod.Lit.*, p. 263.

texts in order to make personal requests.[22] This shift from a political function to the private domain is accompanied by a passage from oral to written expression. Poetry and rhetoric, enlarging their sphere of activity and invading the area of writing, had from now on a deeper impact which is illustrated, for instance, by the language used in documents, even the most common ones, such as letters.

An examination of the work of Dioscorus reveals that the other characteristic of this new culture was the interaction of literature and the document, to take up two concepts used by papyrologists.[23] One could actually speak of a 'documentarisation' of literature. As has been seen with the petition-like poems by Dioscorus, poetry came to serve the writing of documents, and even became a sort of document itself. More generally, the most renowned authors, such as Homer and Menander, obtained a practical role: their writings became as it were handbooks used to compose all sorts of written deeds.

A reverse, but actually complementary, development also took place: the 'literarisation' of the document. The generic boundaries between documentary and literary texts became more and more blurred. Quotations and paraphrases from works by famous authors invaded the documentary genre, and vocabulary derived from poetry was increasingly adopted. Even in daily life one tried to assert one's culture: in writing a letter, a petition or a contract, people attempted to style their texts as works of literature. This phenomenon shows to what extent the society of Late Antiquity was imbued with culture. As a result, the literary quality of documents increased greatly, even if they sometimes fell into a convoluted sophistication. The corpus of Dioscorus, consisting of both literary and documentary texts, exemplifies this phenomenon. Even when he drafts a document, Dioscorus continues to be a poet. That is the case with the majority of his petitions. The petition, a text in which a subject notifies the authorities of a crime of which he has been the victim, and for which he demands justice, often begins at this period with a very elaborate introduction praising the virtues of the official who is addressed. This introductory part resembles an *enkômion* influenced by Greek rhetorical conventions. However, the influence of literature is not restricted to this. It can be observed throughout the text, be it in the quotations or in the vocabulary. I cite only one example, *SB* XIV 11856 (= *P.Berl.Brash.* 19). This is a petition sent by a *colonus* to his patron, probably dictated by Dioscorus or recopied based upon a model written by him:[24]

[22] Cf. *P.Aphrod.Lit.*, p. 269, note 10.
[23] Cf. *P.Aphrod.Lit.*, pp. 684-687.
[24] Cf. Fournet, 'À propos de *SB* XIV 11856'. From this earlier publication I have taken over the text and the remarks with regard to idiom.

† Τῷ ἐμῷ] ἀ̣γαθῷ δεσπότῃ καὶ φιλανθρωπευεργέτ̣[ῃ
]..ον φιλ[ο]πτωχωτάταν μεγαλοπρ[επ
] Δωροθέῳ ἐμῷ γεούχῳ [
]κ̣ο̣ς̣.. λαμπροῦ ὑμῶ[ν
]σ̣ο̣ι̣ω̣ φυλάττοι τὴν ὑ̣μ̣[ε]τέρα[ν].[
λ]α̣μπροτάτων καὶ τριποθή̣[τ]ων α.[ ].ησ[
'μ]έλιτος γλυκίων ῥέεν αὐδὴ' ἐπὶ μήκι[στον χρόνον
]αν ἀναθάλλουσαν τυχεῖν ὅσων ἐρᾶτε α[
τ]ρόποις καὶ καιροῖς· εἰς τί μὴν οὖν, δέσπο[τα,
]μενον καὶ παναγαθοσύνης ἔργον τῆς η[
]ου μεγάλης τοῦ δεσπότου Θ(εο)ῦ καὶ παμβα[σιλέως
]. χεῖραν ὀρέξαι μοι δικαιοσύνης καὶ εὐλ[αβείας
τυγχ]άνω κακούμενος μετὰ τῶν δούλων ὑ[μῶν υἱῶν μου
]ων ἀνακείμενος τυγχάνω δυσεντερικὰ σπ[άσματα
ἀ]λγηδόνας ἀσηχής· διόλου ὑπάρχων μαλι[
'μετὰ] τῶν νηπίων τέκνων ἐφ' ὅτι ἄπρακ[τος
].ικὸς ἐκ χειρῶν ἀπραγείας ἵνα ὁ Θ(εὸ)ς ἀντ[
v° εὐχαρι]σ̣τήσω ὑμῖν διὰ παντός, ἐν πολλοῖς δὲ κα[ιροῖς
]ου †

One discovers:

- a quotation from Homer's *Iliad,* I 249 (l. 7) and a fragment of iambic trimeter (l. 8. τύχειν ὅσων ἐρᾶτε) that can also be found in a poem by Dioscorus;[25]
- poetic or, more generally, literary expressions: χεῖραν ὀρέξαι μοι δικαιοσύνης καὶ εὐλ[αβείας (l. 12) which combines the Homeric phrase χεῖρα ὀρέξαι (cf. *Il.* XXIV 73, and subsequently Euripides, *Phoenissae,* 103) with a qualitative genitive (an artistic device often used in poetry from Late Antiquity); χειρῶν ἀπραγείας (l. 17), another qualitative genitive with particular connotations;
- poetic vocabulary: the epic adjective ἀζηχής (written ἀσηχής), similarly used by Homer to express pain, *Il.* XV 25; the Homeric expression νηπίων τέκνων (*Il.* IV 238, VI 95, 276, 310, XI 113, XVII 223, XXII 63, XXIV 730; *Od.* XII 42, XIV 264, XVII 433);
- rare words with literary connotations: φιλανθρωπευεργέτης (l. 1), [26] παναγαθοσύνη (l. 10).

---

[25] *P.Aphrod.Lit.*, IV, 27,2 (Dioscorus).

[26] A word that one only finds in another petition by Dioscorus, *P.Lond.* V 1677, 2.

The interaction between literature and document reveals several phenomena: the importance accorded to literature and book knowledge, and to poetry in particular, in Late Antiquity; and, most importantly, the fact that in Late Antiquity society came to conceive of writing as unitary: a literary text was no longer necessarily viewed as the opposite of the document, since both forms were subject to the same rules of writing and composition inspired by poetry and rhetoric. This development of a unitary concept of writing, which is so different from our twenty-first-century notions, had enormous consequences: the Byzantines did not distinguish technical learning from literary culture, as is indicated by the fact that the highest posts of administration might not be given to experts in law or civil administration, but to men of letters, and to poets in particular. Thus poets such as Cyrus, Olympiodorus, Claudian or Pamprepius – all descending from Egypt and the Thebaid in particular – became very high officials of the Empire. At a provincial level Dioscorus himself was living proof of this custom: his poetic skills, although elementary, very likely helped him to obtain the post of notary during his stay in the ducal capital of the Thebaid, Antinoopolis. Furthermore, the title of *scholastikos* assigned to Dioscorus (*P.Cair.Masp.* I 67064) is significant in that it implies both literary and legal studies. Such an ambivalence typifies at best this tendency towards non-specialisation, and the confidence placed in literary knowledge by the society of Late Antiquity.[27]

[27] I thank my colleagues James Hirstein and Denyse Vaillancourt for helping me with the final draft of the English translation of this article.

DIOSCORUS AND THE LAW[1]

Peter van Minnen

Dioscorus[2] was a sixth-century notary who drew up a variety of documents for others and for himself. He originated in Aphrodite,[3] a small town in Upper Egypt, but he also made two trips to Constantinople (548/9 and 551) and spent several years in Antinoopolis, the capital of Upper Egypt (565/6-570/3).

As a notary, Dioscorus used his legal knowledge in a very practical way. The documents he composed in Greek and in Coptic followed established patterns, but he also drew on existing law, both imperial and customary. To produce documents in accordance with the latest official requirements, Dioscorus had to be up to date. He may have learned the traditional patterns from other, older texts, which he could imitate and use as precedents. For the latest developments Dioscorus had to know current law, either in the form of translations or through contacts with more experienced colleagues. His visits to Constantinople and his extended stay in Antinoopolis will have much facilitated this.

There was also much scope for rhetorical composition. Some of the documents Dioscorus produced were long prose texts, in which he narrated a case in great detail.[4] For this part of his task other, older texts were available as models. Dioscorus also kept drafts of his own documents in his archive, either for future reference or for educational purposes. These illustrate how he worked and how he may have transmitted his knowledge to others.

Dioscorus is generally known for his prolix style. It is, however, not so much the style of the documents as the exhaustive listing of all kinds of

---

[1] I have profited from discussions with J. Urbanik, Warsaw. All dates are AD unless otherwise noted.

[2] On Dioscorus and his archive see the brief accounts in Bell, 'An Egyptian Village'; Keenan, 'The Aphrodito Papyri'; MacCoull, *Dioscorus of Aphrodito*, pp. 1-56; Verbeeck, 'Dioskoros'; Gagos and Van Minnen, *Settling a Dispute*, pp. 1-48; Kuehn, *Channels of Imperishable Fire*, pp. 42-76; Fournet, *Hellénisme*, vol. 1, pp. 317-343.

[3] Fournet, 'Appendice', points out that in the sixth century the official name of the village was Aphrodite. In early Arab times it was called Aphrodito.

[4] See on this especially the analyses in MacCoull, *Dioscorus of Aphrodito*, pp. 16-56.

provisions and clauses, often very similar from contract to contract, which makes reading them rather heavy going. With the wider dissemination of legal knowledge in Late Antiquity the inclusion of all kinds of provisions and clauses in documents became inevitable. For all the clients knew, without these provisions and clauses, documents could be disputed on some legal point and rendered ineffective. The notaries could charge more for longer documents.[5]

The *pièce de résistance* in Dioscorus's archive are four imperial rescripts (*P.Cair.Masp.* I 67024-67025, 67026-67027, 67028 and 67029).[6] Since these were drafts kept by Dioscorus himself, sometimes in multiple copies with variations from one text to another, it seems to follow that the petitioners were requested by the authorities in Constantinople to draw up their own responses to the petitions they presented to the emperor. This idea, first suggested shortly after the rescripts were published, has been widely accepted by legal scholars.[7]

For this remarkable procedure later European parallels can be adduced.[8] From the second half of the ninth until the twelfth century royal acts could be prepared by the beneficiaries themselves (in German this is called *Empfängerausfertigung*) after which the royal chancery would provide only the finishing touch. It could also take the drafts of the beneficiaries and copy them out in its own style or draw up the acts from scratch (in German this is called *Kanzleiausfertigung*). Even if the royal chancery drew up the acts from scratch, it could still cannibalise the petition presented to the king for the narrative of the facts and for the formulation of his answer. It could also draw on previous acts issued in similar cases (in German these are called *Vorurkunden*). The royal chancery would also have formularies of such texts at its disposal, but this material may rather have been used to train new scribes.[9]

---

[5] Cf. Wolff, 'Der byzantinische Urkundenstil Ägyptens', p. 139, who remarks on 'die namentlich in liebevoll ausgearbeiteten Vertragstexten häufig anzutreffende, ebenso naive wie doktrinäre Sucht ... herkömmliche Klauseln mit praktisch wertlosen, aber Buchwissen verratenden romanistischen Begriffen aufzuputzen'.

[6] The *editio princeps* of *P.Cair.Masp.* I 67026-67027 and 67028 is Maspero, 'Études sur les papyrus d'Aphrodité', pp. 138-152.

[7] See Amelotti and Migliardi Zingale, *Le costituzioni giustinianee*, especially pp. 44-45.

[8] Guyotjeannin, Pycke and Tock, *Diplomatique médiévale*, pp. 228-230.

[9] Lusigna, 'La transmission'. Royal acts were composed like letters, and 'arts épistolaires' in the title of Lusigna's article refers to the royal acts themselves, not to letter books, which, however, also circulated in the Middle Ages.

Dioscorus may well have kept the drafts to instruct trainees. We know of such activity on his part only indirectly.[10] Some texts in his archive exist in two copies, one in his own hand and one in another. Since his own texts were authoritative, and the copies were sometimes mistaken in what they read in his original, it follows that the other texts were written by an associate, who copied such texts for practice purposes. As an example I mention the will of a physician from Antinoopolis (*P.Cair.Masp.* II 67152), which someone copied in *P.Cair.Masp.* II 67151. The copy reproduces Dioscorus's abbreviation of *ny* at the end of a line (by a supralinear stroke) in the middle of lines, where it is inappropriate. Another example is provided by some of the semi-literary texts in Dioscorus's archive. A metrological table in his hand (*P.Lond.* V 1718) was copied in an unpublished text[11] in a hand which also wrote the conjugation tables in the archive (*P.Aphrod.Lit.* III 2-4).[12] These tables seem rather out of character for an accomplished scribe such as Dioscorus. The person who wrote them may well have been his trainee. The *ethopoiiai* included among Dioscorus's poems (*P.Aphrod.Lit.* IV 41-46), which work out set themes for composition, may also point in the direction of his role as educator.[13] The Homeric 'glossary' (*P.Aphrod.Lit.* II), the life of Isocrates (*P.Cair.Masp.* II 67175) and the bilingual glossary[14] might also belong here, but it seems to go too far to regard Dioscorus as a full-blown *grammatikos*.

Dioscorus kept several drafts of imperial rescripts in his archive. We do not know whether the final versions handed in to the imperial chancery were copied out in its own style or whether it finished them off by merely having the emperor add the imperial subscription and by adding the 'seen' note vouching for the authenticity of the subscription. The final texts would in any case have been issued by the imperial chancery. Still, the fact that petitioners from Aphrodite including Dioscorus himself were requested to draw up the responses shows how legal nullities – as Dioscorus and his companions no doubt were as compared to the legal experts in Constantinople were somehow involved in the creation of law in late Antiquity. One might argue that the rescripts in this case are merely referrals from the emperor to the local authorities in Upper Egypt and that more serious rescripts would have been drafted by the imperial chancery itself.[15] But even these referrals required exact legal formulation. If this could be left to people like

[10] Fournet, *Hellénisme*, vol. 1, pp. 236-237 and 325-326; vol. 2, pp. 688-690.

[11] To be published by J.-L. Fournet.

[12] Cf. similar tables in *P.Aphrod.Lit.* III 1.

[13] Fournet, *Hellénisme*, vol. 1, pp. 275-276.

[14] Hasitzka, *Neue Texte*, no. 257.

[15] Amelotti and Migliardi Zingale, *Le costituzioni giustinianee*, pp. 44-45.

Dioscorus, the imperial chancery may well have taken a chance also with more complicated cases. One is tempted to compare this reliance on the legal know-how of members of the public with the responsibilities with which *quaestores* were credited in Late Antiquity.[16] These magistrates were not always legal experts themselves, but they were nevertheless called upon to draft laws. The *quaestores* and Dioscorus and his companions, who in their own little way were also involved in the creation of law in late antiquity, shared the same basic literary education, which allowed them to be trusted with it in the first place.

Dioscorus's activity as a notary dates from after the promulgation of Justinian's legal corpus.[17] This means that he could not ignore existing legislation. The problem was how to apply the law to each case and even where to find it. Although the Justinian legal corpus is organised by topic, it is by no means apparent what the decisive piece of legislation is in each case. Since the material is arranged chronologically under the various headings, it would perhaps make sense to quote only the most recent law, much in the way in which modern scholars tend to refer to the latest scholarly contribution rather than to the whole series of contributions or even the most pertinent contribution – which, in fact, is not always the most recent. In the case of late Roman legislation, the situation was even more complicated. Newer laws did tend to abrogate older laws, but if older laws contained more details, which could still be relevant, they would still be valid. If the latest law on a particular topic did not contain the necessary details, one would have to find one's way back through the older laws to find out what provided the best reference for a particular case.

The main reason why Dioscorus went to Constantinople in 548/9 and again in 551 was to obtain a rescript on behalf of his hometown. For the first visit we have a draft of a rescript (*P.Cair.Masp.* I 67029), which is unfortunately much damaged. It is not in Dioscorus's hand,[18] but presumably in that of one of his companions. For the second visit we have two much better preserved drafts of a rescript (*P.Cair.Masp.* I 67024-67025). On the first draft, which is in Dioscorus's hand,[19] he rewrote a section of the text on the back (lines 30-52) below an isopsephistic poem on Saint Menas (*P.Aphrod.Lit.* IV 48). The easiest explanation for this is that the rescript

[16] See Honoré, 'Ausonius and Vulgar Law'.

[17] Dioscorus first occurs in 543 (*P.Cair.Masp.* I 67087). The earliest text he wrote is *P.Cair.Masp.* III 67283 of 547/8.

[18] For help in sorting out the hands responsible for writing the texts mentioned in this paper I am indebted to J.-L. Fournet, who checked his set of photographs for me.

[19] The second draft is not in Dioscorus's hand, but in that of one of the other drafts. I will return to this later on.

was drafted in multiple stages and that Dioscorus copied the isopsephistic poem while making the first draft. Since there are no other isopsephistic poems among Dioscorus's papers, we may assume that he merely copied this text from someone else and that he did not compose it himself. He could find a lot of interesting poems in Constantinople. Saint Menas happened to be an Egyptian saint of international fame.

The portion of the rescript on behalf of Dioscorus's hometown copied below the isopsephistic poem represents an intermediate stage between the two larger versions, which strongly suggests that it was written on the papyrus while he was still in Constantinople. That Dioscorus rewrote this particular portion of the rescript by copying it sentence by sentence from his draft on the front seems indicated by the letters (all *alpha*s) found between the lines on the front of the papyrus. These are not found in other texts in Dioscorus's archive. That they denote a particular type of prose rhythm, as has been suggested,[20] seems unlikely. That they have to do with the various stages of the preparation of the drafts of the rescript seems more likely. While copying the text from the front Dioscorus may have marked for himself how far he had got while flipping from one side of the papyrus to another.

All three versions of the rescript on behalf of Dioscorus's hometown, where they exist, show slight differences. That these texts are drafts and not copies also follows from the fact that the prescript is incomplete: 'X (name left out) from such-and-such a village (ἐκ τῆσδε τῆς κώμης)' in lines 1-2.[21] The final draft would have provided proper identification and would also have included a concluding stipulation about a fine. The other rescripts drafted during Dioscorus's second visit to Constantinople are also to some extent incomplete. The rescript he drafted on behalf of his hometown is extremely interesting, but I will concentrate on two shorter rescripts which deal with private legal matters, either concerning Dioscorus himself or his family.

At this point it is important to work out the family relationship implied by the two shorter rescripts. The editor thought that the two texts (*P.Cair.Masp.* I 67026-67027 and 67028) were exercises, because in his view they were incompatible.[22] Both seem to relate to the same person, Dioscorus, but the editor could not reconcile the facts about this person contained in the two texts. Later scholars, commenting on the texts from a legal

[20] Wendland *apud* Partsch, 'Neue Urkunden', p. 205, note on lines 39-43.

[21] The use of the demonstrative pronoun should not be taken as an indication that the text was written in Aphrodite. The demonstrative is used in a similarly generalising way in *P.Cair.Masp.* I 67027.

[22] Maspero, 'Études', pp. 149-152.

point of view, have claimed that the objections are indecisive.[23] One scholar, however, noticed that the data in the texts were not easily reconcilable with what was otherwise known about the family of Dioscorus.[24] As an alternative to the idea that the texts were merely exercises, he thought the texts could relate to someone else, either called Dioscorus or not. In the latter case Dioscorus would have substituted his name for another.

When I first read these texts I noticed a problem, which, oddly enough, had not been remarked upon by the legal scholars. In *P.Cair.Masp.* I 67028, which is written in Dioscorus's hand, he tries to secure his mother's inheritance, which had been appropriated by his half-brother(s) and/or -sister(s). He clearly acts alone and therefore does not have another full brother or sister who could also claim part of his mother's inheritance. In *P.Cair.Masp.* I 67026-67027, however, Dioscorus does have a full sister, who is represented by her brother. My initial conclusion was that the two Dioscori were not one and the same person.

In a recent monograph[25] I reconstructed the family of Dioscorus a little further than previous scholars. I included two cousins, children of a sister of Dioscorus's father, called Anastasia alias Tekrompia and another Dioscorus. It now appears that *P.Cair.Masp.* I 67026-67027 concerns them, not our man.[26] The other Dioscorus and his sister had been entrusted to the care of an uncle called Apollos, who had recently died. This must be Dioscorus's father. The sister is called Peristera, 'the dove', in the second draft (*P.Cair.Masp.* I 67027), which is in Dioscorus's hand. The Coptic alias of his cousin, Tekrompia, also means 'the dove'. In the second draft Dioscorus apparently used the Greek rendering of this alias to identify his cousin.

The first draft (*P.Cair.Masp.* I 67026) is in the same hand as the second draft of the rescript on behalf of Dioscorus's hometown (*P.Cair.Masp.* I 67025). It may well have been written by Dioscorus's cousin Dioscorus, who accompanied him on his second visit to Constantinople. Apparently Dioscorus drafted the rescript on behalf of his hometown (*P.Cair.Masp.* I 67024) and on his own behalf (*P.Cair.Masp.* I 67028), but submitted only the former to one of his companions for a 'second opinion' (*P.Cair.Masp.* I 67025), presumably his own cousin Dioscorus. This Dioscorus must also have been quite knowledgeable about the law, if he worked on the rescript on behalf of his hometown and made the first draft of the rescript on behalf

---

[23] Partsch, 'Neue Urkunden', pp. 228-229 and 232-233, and Amelotti and Migliardi Zingale, *Le costituzioni giustinianee*, p. 57.
[24] Bell, 'An Egyptian Village', pp. 26-27, note 24.
[25] Gagos and Van Minnen, *Settling a Dispute*, p. 131.
[26] This has now also been argued by C. Zuckerman at the International Congress of Byzantine Studies in Paris (August, 2001).

of himself and his sister (*P.Cair.Masp.* I 67026).[27] He had of course received the same kind of education as Dioscorus.

If *P.Cair.Masp.* I 67026-67027 does not concern Dioscorus personally, although he wrote the second draft, but another Dioscorus, does *P.Cair.Masp.* I 67028 really concern our man? Or could this be a rescript drafted by Dioscorus for yet another Dioscorus? This seems hardly credible. In this text Dioscorus is the only surviving child of his father and mother. He does not mention other heirs to his mother's inheritance, but he does mention half-brother(s) and/or -sister(s), children of his father and his second wife who did not share in his mother's inheritance. We know that Dioscorus indeed had a brother, Menas, and a sister, who may well have been his half-brother and -sister. As children of his father, but not of his mother, they shared in the inheritance of his father only. This is how they appear in other documents.

One receipt for taxes on land formerly owned by Dioscorus's father Apollos is made out in the name of Menas 'his son' (*P.Lond.* V 1702). Similar receipts exist made out in Dioscorus's name, so we may assume Menas was a half-brother, a child of the same father. In another text Dioscorus seems to have been on bad terms with Menas, because one of his fields has been damaged by two *protokometai* at the instigation of Menas 'my brother' (*P.Cair.Masp.* III 67319, 16). An anonymous sister occurs in *P.Lond.* V 1677, where her husband and Dioscorus's son have both been maltreated by the pagarch, although, as Dioscorus points out, his son had no business with that husband. Again we may assume that she was a half-sister. One scholar[28] has detected another sister in the wife of Phoebammon who appears as a son-in-law of Apollos (his name appears in a lacuna in *P.Cair.Masp.* I 67108, 8) and as a representative, along with Dioscorus, of Apollos's heirs. We now know that Phoebammon was married to Anastasia 'the dove', a niece of Apollos.

In the rescript on behalf of his hometown (*P.Cair.Masp.* I 67024-67025) a special section (lines 24-30) is devoted to the inheritance of Dioscorus's father, which had been seized by others.[29] This problem is referred to in passing in the rescript drafted for his cousins (*P.Cair.Masp.* I 67026-67027). *P.Cair.Masp.* I 67024-67025 mentions both Dioscorus and his half-brother and -sister as the duped party, which is exactly what we should expect if they had the same father and shared his inheritance. When Apollos's

[27] I have not yet been able to compare the hand of *P.Cair.Masp.* I 67025 and 67026 with the only known signature of Dioscorus's cousin Dioscorus in *P.Cair.Masp.* III 67283 iii 7-8.

[28] Bell, 'An Egyptian Village', p. 26.

[29] I have translated this passage below in Appendix 1.

property was seized after his death, all his heirs, not just Dioscorus, would have been duped. While the culprits in *P.Cair.Masp.* I 67026-67027 are identified as men of Julian the pagarch (line 7), though their claim on Apollos's property is not specified, it is clear from *P.Cair.Masp.* I 67024, 24-30 that we are dealing with the same people there. The crossed-out phrase (line 27) refers to a tax claim on Apollo's property, which makes good sense if the culprits were acting on the pagarch's behalf. Earlier on in this text it appears that Dioscorus had gone to Constantinople for this once before, which must be the visit of 548/9. For this earlier visit the petition *P.Cair.Masp.* I 67019, which identifies the culprit as the pagarch, and the rescript *P.Cair.Masp.* I 67029 were drafted. The somewhat earlier report *P.Cair.Masp.* III 67283 also identifies the culprit as the pagarch. During the earlier visit Dioscorus secured a letter (*SB* VI 9102) in which the pagarch is again the culprit and the duped party is described as οἱ παῖδες Ἀπολλῶ (lines 3-4), i.e. Dioscorus and his half-brother and -sister.[30]

Dioscorus grew up as the only child of his father and mother. After the death of his mother, his father remarried and had at least another son and a daughter, with whom Dioscorus grew up as the elder half-brother. In addition, the house of Apollos was occupied by two cousins of Dioscorus, a boy and a girl, whose mother had died while they were still very young (their father had presumably died even earlier). Dioscorus seems to have had a good relationship with these cousins, because he helped draft the rescript on their behalf (he wrote *P.Cair.Masp.* I 67027). In the other rescript where he is defending his own rights (*P.Cair.Masp.* I 67028), the children of his father's second marriage are shown in a much less charitable light, although he cannot fail to mention them in tandem with himself in the section of the rescript on behalf of his hometown dealing with the inheritance of his father (*P.Cair.Masp.* I 67024, 24-30). The irritation over his mother's inheritance could be reflected in Dioscorus's later suspicion that his half-brother was behind some of the damage done to his fields. But perhaps I am reading too much into this.

This leaves me with just one nut to crack, another brother who has been claimed for Dioscorus by earlier scholars. Senuthes (or Sinuthios) son of Apollos has been identified as his brother by the editor of *P.Hamb.* III 231, whose idea has been widely accepted.[31] As in many other texts, Senuthes appears there as a *protokometes* alongside Apollos, Dioscorus's father. The editor believes that her text actually proves that Senuthes was Apollos's son, but in fact it proves quite the opposite. In the text someone acknowledges the receipt of Dioscorus's third share of a debt once incurred by

---

[30] A revised stemma of Dioscorus's immediate family appears below as Appendix 2.

[31] Most recently by Koenen in Gagos and Bagnall, *P.Thomas*, p. 236.

Apollos and Senuthes. On the back of the text Dioscorus identifies Apollos as his father, but Senuthes he objectively identifies as 'Senuthes the son of Apollos'. If Senuthes had been the son of his father and not of some other Apollos, Dioscorus would have made that explicit. As it stands, Senuthes must be the other *protokometes*, a colleague of Apollos, but not his son. Senuthes survived Apollos, and Dioscorus took over his father's position as *protokometes* alongside Senuthes. Whether Senuthes was related to Dioscorus in another way, we cannot tell, because the names are all very common. Apollos and Senuthes sometimes acted together with Victor, the priest of the main church of Aphrodite, who was another cousin of Dioscorus. In *P.Mich.* XIII 669 the three of them borrow money. There appear to have been three *protokometai* in Aphrodite.[32]

It is now time to look at the texts in more detail. In *P.Cair.Masp.* I 67028[33] Dioscorus refers to a law of Leo of 472, which we know as *C.* 5, 9, 6 with other bits and pieces in *C.* 6, 20, 17 and 6, 61, 4, on the constraints on a man's property in case he remarries. Dioscorus quotes a part of this law below the draft of the rescript and refers to it in lines 11-15. Properly speaking, this particular point is not the reason for his complaint against his half-brother and -sister in lines 1-11. That is not the property of his father, but that of his first wife, Dioscorus's mother. His father had remarried and transferred to the second wife what he had previously given to his first wife out of his own property without regard for the rights of his son from the first wife.

It is very likely that Dioscorus's father had remarried because his first wife had died. This would seem to be indicated by the quotation of the law of Leo which follows below the text of the rescript, which envisages a husband who survives his first but not his second wife, but since this seems to be merely a fragment of a much more complicated piece of legislation, we do not have to apply it literally to the case at hand. After Dioscorus's father died intestate, the children of the second wife had appropriated the lot originally belonging to Dioscorus's mother without any regard for his rights. His father's second wife had presumably also died in the meantime, because she does not play an active role in the matter.

In the translation I have taken τοὺς ἐξ ἐκείνης παῖδας in lines 4-5 as the subject of the infinitive. Many commentators seem to have assumed that

[32] Fournet, 'Le système des intermédiaires', p. 240, points out that *P.Cair.Masp.* III 67286 identifies all three as *protokometai*, but strictly speaking the priest and the two *protokometai* merely represent the *protokometai* in that text. But the point is no doubt a valid one. Fournet does not include *P.Lond.* V 1702, which is dated to a sixth indiction (557/8?).

[33] A translation of this text is included in Appendix 1.

this expression is the object of the infinitive.[34] The verb ἀφελέσθαι, however, does not mean 'to disinherit' but 'to take away' (as in the other rescripts, *P.Cair.Masp.* I 67024, 26 and 67026, 8), and the object of the verb is no doubt the same as that of καταδαπανᾶν: the things which originally belonged to the first wife, Dioscorus's mother. Dioscorus wants to make sure that he gets this back. At this point he does not invoke the law of Leo, which deals with the constraints on a man's property should he remarry. In lines 1-11 of the text Dioscorus deals only with his mother's property. He claims all of it. His father had no right to transfer this to anyone, least of all to the second wife. Dioscorus has been deprived of the inheritance which he as the only child of his mother – the legal and only heir to the property concerned with in lines 1-11 – could naturally claim as his due. The use of ἔρημον in line 6 also points to his being deprived of his due rather than to his being an orphan.

I cannot help remarking that the editor saw no difficulty in lines 1-11,[35] where he nevertheless mistranslated a vital passage, and that the subsequent legal commentators have gone so far as to claim that the legal case as a whole was crystal clear.[36] On the basis of lines 11-15 they thought that Dioscorus's father even disinherited the children of the first wife, which is contradicted by Dioscorus's claim to a rightful share in his father's inheritance. The editor thought that Apollos had merely favoured the children from the second wife in his will. If there had been a will disinheriting Dioscorus, he should have disputed it, but this he does not do. I rather think that the inheritance of Dioscorus's father had not yet been divided and that there was no will.

In lines 11-15 Dioscorus merely mentions, just to be on the safe side, an additional point which affects his continuing relationship with his half-brother and -sister. With regard to his father's own property they naturally shared in the inheritance. Dioscorus cannot well claim all of it. In this connection he invokes the law of Leo, which stipulates that the second wife receives only as much as each child from the first marriage. As the only child

---

[34]See the translation in Maspero, 'Études', p. 142 as explained at 147. Only Beaucamp, *Le statut de la femme*, p. 69, construes it correctly.

[35] Maspero, 'Études', p. 146.

[36] Partsch, 'Neue Urkunden', p. 231: 'Der Rechtsfall macht dem Verständnis keine Schwierigkeit'. Mitteis in the introduction to his *Grundzüge*, no. 382: 'Der Rechtsfall bedarf keiner Erläuterung'. At least Maspero was aware of problems in the rest of the text ('Études', p. 146): 'La seconde affaire est loin d'être ... intelligible, il s'en faut de beaucoup'), but did not realise what the decisive problem was (the absence of a full sister *P.Cair.Masp.* I 67028, the presence of one in *P.Cair.Masp.* I 67026-67027). Beaucamp, *Le statut de la femme* 2, pp. 68-70, did not realise it either.

from that marriage, Dioscorus would get at least as much as the second wife or, now that she is apparently dead, her heirs. In the absence of a will he had a right to be worried about the share of his father's inheritance that he would end up with.

An additional reason why my interpretation of τοὺς ἐξ ἐκείνης παῖδας in lines 4-5 as the subject of the following verbs is to be preferred, is the fact that otherwise Dioscorus would be using a plural to refer to himself, whereas no other heirs from the first marriage are mentioned as potential beneficiaries in what follows. He clearly was his mother's only child. Of course, one can always claim that Dioscorus was a muddled thinker, but my experience in dealing with his texts is that they are usually very precise. It is just careless reading which has not always shown Dioscorus in the best of lights. Yet his documents have often been referred to in contexts where a muddled understanding of them is potentially misleading. In other cases (very) important information contained in these texts has simply been disregarded, because no one has understood the texts for lack of a careful translation. This situation should no doubt be addressed first before more monographs on Dioscorus are produced.[37]

The matter of his father's inheritance (not that of his mother's, which is the main point dealt with in lines 1-11) is further complicated by the fact that his father had received a donation from an aunt. The children from the second marriage had made this aunt write a new donation document, in which they, rather than their father, were made the beneficiaries of the donation. This would prevent Dioscorus from claiming any part of this donation, because it would not be part of his father's inheritance in which all his children would share. Lines 15-22 are therefore devoted to pointing out that the aunt's original donation should remain part of his father's inheritance. Generally speaking a second donation does not abolish the rights established by a first donation. This means that it was the aunt, not Apollos, who was made to write a second donation. As long as Apollos was alive, there would seem to be no particular reason to transfer the donation from him to his children from the second marriage only. This would be up to him. Presumably, Dioscorus's father was already dead by the time the second donation was written. The children from the second marriage would not have bothered to have the aunt write a new donation in their favour, if they did not find themselves with Dioscorus on their (opposite) side as co-heir of their father's estate of which the aunt's donation had become part.

Second donation documents were legally invalid except in circumscribed cases, e.g. when someone wanted to disinherit a child for bad behaviour towards his parent after the child had been the beneficiary of a do-

[37] Cf. the recent blossoming of such monographs mentioned in note 2.

nation document. The way was then open for a second donation document regarding the same property. Here the matter seems to be as Dios-corus suggests: the children of the second marriage try to extract as much as possible out of their father's inheritance before dividing it with Dioscorus. The second donation document is not only illegal, but really a devious way to harm the rights of their co-heir. The imperial legislation against second donations had come into being to prevent just this type of fraud.

Whether Dioscorus is telling the truth or withholding important information, we cannot tell. The present rescript is merely a way to open the case and to have the authorities in Upper Egypt look into the matter. If Dioscorus was really as poor and deprived as he makes himself out to be, he could only stand to gain by having the matter looked into. He would get his mother's inheritance *in toto* and at least a share of the aunt's donation. As far as the general division of his father's inheritance is concerned, Dioscorus would need all the support he could get to secure his rightful share of it. He could not well have dropped the reference to the law of Leo, because it supported his case.

What Dioscorus added below the text of the rescript does not correspond to anything in the Latin text of the law of Leo preserved in Justinian's *Code*. As one of the early legal commentators already saw,[38] Dioscorus must have had access to a Greek translation of the complete law, not just of the excerpt we find in the *Code*. The way *C.* 5, 9, 6 opens shows that the excerptors left out the *considerans*, the narrative leading up to the actual decision of the emperor. In the Latin text *sancimus* corresponds to θεσπίζομεν in Greek, which we find several times in the drafts kept by Dioscorus (*P.Cair.Masp.* I 67024, 17, 27 and 39; 67026, 12; and 67028, 7 and 18). There it is always preceded by a substantial *considerans*, and this was no doubt also the case with the law of Leo. What we find below the rescript is presumably the original opening of this law.

In Dioscorus's draft the emperor urges the *dux* of Upper Egypt, the addressee of all the rescripts Dioscorus and his companions were after in Constantinople, to uphold the law of Leo. In fact, in Justinian's legal corpus we find similarly worded references to this law in legislation issued by Justinian himself and excerpted in *C.* 5, 9, 10 of 529 and *Nov.* 22, 27 of 536. The law of Leo must have been the key text which both issuers and users of the law in late antiquity referred to when it came to inheritance problems caused by second marriages (which were disapproved of for religious reasons).

P.Cair.Masp. I 67028 shows that Dioscorus was able to imitate the style of imperial rescripts and to quote a law of Leo. Dioscorus may have

---

[38] Partsch, 'Neue Urkunden', p. 234.

been pointed to this by someone in Constantinople, but maybe he already knew it. To decide this, we have to consider whether the rescripts were drafted in Aphrodite or in Constantinople. One of the drafts made for the earlier visit to Constantinople in 548/9, a petition (*P.Cair.Masp.* I 67019), is written on the back of an unrelated text. This draft could have been written while Dioscorus was still at home in Aphrodite. Since the petition concerns Dioscorus's hometown, it could have been shown to some of his fellow villagers before departing for Constantinople. The rescript drafted subsequently (*P.Cair.Masp.* I 67029) was written on a clean sheet of papyrus. One of Dioscorus's companions may well have written it while in Constantinople.

The petition (*P.Cair.Masp.* I 67019) also refers to a constitution (a τύπος) of Leo, but in this case one concerning Dioscorus's hometown directly. This text and its confirmation by Justinian, which is also mentioned, would have been known and available in Aphrodite itself. In the case of the rescript which Dioscorus intended to procure for himself in Constantinople in 551 (*P.Cair.Masp.* I 67028) there would not seem to be a particular reason why he should have drafted this text already in Aphrodite. It was written on a clean sheet of papyrus, just as the other rescripts drafted for his second visit to Constantinople (*P.Cair.Masp.* I 67024-67025 and 67026-67027). Dioscorus could therefore have found the Greek text of the law of Leo concerning the inheritance of a man with two wives in succession in Constantinople itself while drafting the rescript.

Would the imperial chancery have told him about the law of Leo? This is not very likely. The text quoted is not to be found as such in Justinian's legal corpus, and the imperial chancery would have been the last to quote an outdated text. But since *P.Cair.Masp.* I 67028 is merely a draft one cannot quite exclude it. On the other hand, if the imperial chancery instructed Dioscorus on how to draft a rescript, there would have been no reason to have him quote the law explicitly, since they would surely recognise the reference in lines 11-15. It is possible that they instructed him to draft a rescript and to refer to the proper legislation and that Dioscorus then set out to get hold of the Greek text of the law of Leo and to draft the rescript. In the draft he both included a reference to the law and quoted its opening.

Why did Dioscorus quote the law explicitly? Those receiving the draft would be able to identify the reference and did not need the quotation. Dioscorus is here quoting from an older translation rather than from the latest text as promulgated in Justinian's *Code*. Strictly speaking one was no longer allowed to quote outdated texts any more.[39] It seems that Dioscorus

[39] Partsch, 'Neue Urkunden', p. 234, and Amelotti and Migliardi Zingale, *Le costituzioni giustinianee*, p. 57.

added the opening merely for himself, which means that he either knew the rest of the text of the old law by heart or that he owned a copy of it (the opening would allow him to find it back among his own papers). On balance I much prefer the latter.

That we have not found any legislation among the papers of Dioscorus can be explained by the fact that only that which no longer interested his heirs was deposited in the storage where his papers were eventually found. Juridical texts and also biblical texts would have been removed by whoever went through Dioscorus's papers after his death, but originally they may have been a not unimportant part of it. *P.Aphrod.Lit.* is not only important in what it contains, because that does give us an idea of the kind of things Dioscorus was interested in, but also in what it does not contain, because that kind of literature would have interested his heirs, i.e. juridical and biblical texts. Dioscorus's interests were apparently larger than those of his heirs, and they included biblical and juridical texts now lost to us. If the interests of his heirs were more restricted, they might have taken just the juridical texts, leaving the biblical texts for the monastery founded by Dioscorus's father[40] and depositing the rest in a storage bin to be retrieved only in modern times.

The second rescript drafted by what appears to be Dioscorus's cousin and by Dioscorus himself (*P.Cair.Masp.* I 67026-67027) concerns the cousin and his sister. She and her brother are known from other texts as the children of Megas and an anonymous mother, a sister of Dioscorus's father Apollos. They grew up with Dioscorus in his father's house. The cousin accompanied Dioscorus to Constantinople to present a petition on his own and his sister's behalf. Dioscorus helped draft the rescript for them, because the second draft is in his hand. The cousin is one of the signatories to *P.Cair.Masp.* III 67283, where his signature (iii 7-8) was identified as being in Dioscorus's hand by the editor. It is not stated there that Dioscorus wrote for his cousin, and it is very likely that the hands of the two cousins were very similar. They grew up together and certainly received the same kind of education. I have, however, not yet been able to check the hand of the signature against the other texts (see note 27).

---

[40] It is often stated in the secondary literature that Apollos at one point became a monk, but this does not appear to be the case. He remained *protokometes* of Aphrodite until he died. In *PSI* VIII 933 he is styled 'apa,' but that does not mean he was a monk. The adjective σεβάσμιος also used for Apollos in that text refers to his status as a *protokometes*. The man addressed in *P.Cair.Masp.* I 67064 may not be Dioscorus's father (see note 44) and does not have to be a monk in a monastic community, as is sometimes claimed, but could be a paterfamilias with an extended family.

What strikes me in the second rescript is that between the lines it remarks on Dioscorus's own situation. The uncle of the cousins is after all his father. People acting for the pagarch have seized Apollos's property because of outstanding debts. The drafts refer to this twice and express strong reservations about this procedure by first using the verb φάσκοντας in line 9 and then λέγουσιν in line 17 to suggest that the authorities had no right to act on their (mere) claims. Although the proper subject of the rescript is the plight of Dioscorus's cousins, his own plight also shines through the text.

Apart from the legal problems with Dioscorus's own inheritance and that of his cousins, his services were also called upon by others, especially after he moved to Antinoopolis. There he composed *P.Cair.Masp.* III 67353 in 569, which is preserved in fragments only.[41] In this *exheredatio* a father disinherits his four children granting them the 'Falcidian portion' only. The old *Lex Falcidia* of 40 BC prescribed that the instituted heir(s) should get at least one quarter of the assets. Here, however, we are not dealing with a quarter but with a third, because the four children are each to get one twelfth (lines 14-15). This is clearly not the 'Falcidian portion', but Justinian's 'legitimate portion', which prescribed that children – who were bound to dispute a will if it entirely disinherited them – should receive at least a third (or one half where there were more than four children) of what they would get in case of intestacy. Is this confusion of terms just 'legal psychology', an example of Dioscorus's inappropriate use of a juridical term to impress his clients? I do not think so. Justinian himself implicitly refers to the 'Falcidian portion' when introducing the 'legitimate portion' in *Nov.* 18, 1, where he raises the amount from a quarter to a third (or one half). He clearly thought he was merely revising the old 'Falcidian portion'.

Just as in the case of the law of Leo, which was the basic reference for issuers and users of the law in cases of a second marriage, the old *Lex Falcidia* was in people's minds when applying the new ruling of Justinian. That emperor had merely changed the details of, but not the idea behind, the old law. The 'Falcidian portion' was late ancient shorthand for complex legislation which culminated in *Nov.* 18, 1. When applying its provisions one could apparently still refer to it as the 'Falcidian portion' – everyone 'in the know' would understand what was meant. This is therefore not a mistake on Dioscorus's part, but rather shows that he was juridically up to date.

In conclusion, it may be added that Dioscorus not only availed himself of his legal knowledge to defend his rights. He also supported his own case by writing 'petition' poems, or even merely laudatory poems, for people in power from whom he expected help.[42] Whether Dioscorus ever used magic

---

[41] For an English translation see Hunt and Edgar, *Select Papyri*, no. 87.

[42] See Fournet, *Hellénisme*, vol. 1, pp. 317-325.

to support his own case is not certain. The only piece of magic included in his papers is concerned with the inundation and shows him rather as a worried landowner.[43]

At the time he embarked on a career as a simple notary,[44] Dioscorus had received whatever legal training was to be had in Aphrodite or nearby Antaeopolis. The complicated structure of his own family, which I have tried to elucidate, provided a better legal training. After the death of his father Dioscorus was forced to secure a large part of the family inheritance for himself and for his half-brother and -sister (*P.Cair.Masp.* I 67024) and even for his cousins. In doing so he availed himself of existing legislation, such as that of the emperor Leo, which he knew through a Greek translation. Later on, when he was an established notary in Antinoopolis – he had asked to be allowed to set up shop as a *nomikos* there in one of the 'petition' poems (*P.Aphrod.Lit.* IV 12) – he continued to practise his legal knowledge, not only for his own sake, but also for others. The always tenuous transmission of property from one generation to another is illustrated not only by the mass of extant legislation on inheritance from late antiquity, but also by Dioscorus's use of this legislation in his attempt to defend his rights against his half-brother and -sister (*P.Cair.Masp.* I 67028) and in his concomitant attempt to help defend the rights of his cousins against authorities who were seizing his father's inheritance while lumping their inheritance incorrectly but understandably with that of himself and his half-brother and -sister (*P.Cair.Masp.* I 67027). The precise composition of Dioscorus's family eluded the authorities just as it has eluded us for the past ninety-five years.[45] I hope to have cleared up some of the muddle.

---

[43] Preisendanz, *Papyri graecae magicae*, no. 13a.

[44] In a letter to what is basically an anonymous father (the name read in the address is largely in a lacuna), someone calls his son Dioscorus a *scholastikos* (*P.Cair.Masp.* I 67064, 13). This title would be out of place for our man. Cf. note 40.

[45] The first text from the archive (*P.Flor.* I 93) was published as early as 1906.

Appendix 1: translations of the more important texts referred to above

P.Cair.Masp. I 67024-67025, 24-30:
Because he (Dioscorus) also says that some of those who own land at that village (Aphrodite) have seized property belonging to the petitioner and his (half-) brother(s) and/or (-)sister(s) in contradiction with what is lawful (crossed out: using the second levy of the aforesaid tax as an excuse), we rule that your illustriousness who watches after their interests should make sure that, if you find out that it is so, justice is done to the petitioner and his (half-)brother(s) and/or (-)sister(s) according to the law.

P.Cair.Masp. I 67026-67027:
Dioscorus approached us saying that his own mother, while the things which had been left to them out of their parents' (estate) were undivided between her and her brother Apollos, passed away with the suppliant and his sister, her children, as heirs [5] while they were still very young, leaving the suppliant and his sister in the care of the aforesaid uncle (*P.Cair.Masp.* I 67027 adds: that this man [Apollos] a certain time before his death handed over to Dioscorus and Peristera, our suppliants, the things and the increments which pertained to them out of the maternal inheritance of so-and-so [the precise reference of this masculine form is unclear], for which they have been paying the taxes until now); and that, now that the aforesaid Apollos has died, the men of Julian the most illustrious (pagarch) proceeded (*P.Cair.Masp.* I 67027 adds: against the things of our suppliants) and took away all the property left by him, claiming that the aforementioned Apollos [10] died with a liability to them, although the maternal things of the petitioners were among that property and not subject to the debts of that man (Apollos). He (Dioscorus) requested to receive our help in this matter.

We therefore rule that your illustriousness should summon both parties and investigate what has been said by the suppliant, and that you according to the laws uphold for him and his sister [15] the ownership of their maternal things according to the share which pertained to her (their mother; not: them, as in Maspero's translation) of the property of her parents together with all legal increment and that you do not allow those who say that the property of the uncle of the suppliants is liable (to them) to subsume also the things which pertain to the suppliants out of their mother's (estate). For even if on the one hand there never was a division between [20] them (their mother and their uncle Apollos), yet on the other hand we do not want (note that the verb βουλόμεθα in line 21 does not properly belong with the expression καθ' οἱονοῦν τρόπον, with which it is nevertheless joint in such a way that one would almost take it as the verb of a subordinate clause) that our petitioners are harmed in any way.

Your illustriousness and those who will hold your position after you and the bureau under your command shall pursue this until the very end by all means, there being a fine of three pounds of gold in store for those who dare to break this ruling or [25] allow this ruling to be broken.

P.Cair.Masp. I 67028:

Dioscorus approached us saying that his own father ... the most prominent ... concluded a second marriage and transferred what belonged to the first wife to the second wife; that he passed away so that the [5] children of (the second wife) took away (not: so that he deprived the children of [the first wife], as in Maspero's translation) and wasted those things; and that our suppliant is altogether laid waste so that he is in need of our protection and that of the laws.

We therefore rule that your illustriousness should summon them and inquire after this; and that, if you find out that it is so and that the things turn out really to have belonged to his [10] mother (supplement uncertain), you should demand this back and give it to the petitioner, while upholding the law of the late Leo for the petitioner with regard to the other things which belonged to the deceased to the effect that the second wife does not receive more than what pertains to each individual child from the first marriage.

[15] Because he also says that some landed property devolved on his own father, after his (whose? Apollos's or Dioscorus's?) aunt had written a donation document, and that the children from the second marriage keep hold of this property (reading τ̣α̣ῦ̣τ̣[α] with the *editio princeps*) after having procured for themselves a second donation document in their name, we rule that your illustriousness should also inquire after this and if you find out that it is so, [20] do not pay any attention (reading προσ⟨σ⟩χῇς) to second donation documents written about things which have already been alienated, considering that even our laws do not make such things valid (tentatively translating Schubart's suggestion ὡ]ς οὐδέ).[46]

The law of Leo is as follows: As much as the first wife would get [25] if her husband dies before her, that much the second wife (should) also get if her husband dies before her (?).

---

[46]See Bilabel, *Berichtigungsliste*, p. 39.

Appendix 2: revised stemma of Dioscorus's immediate family

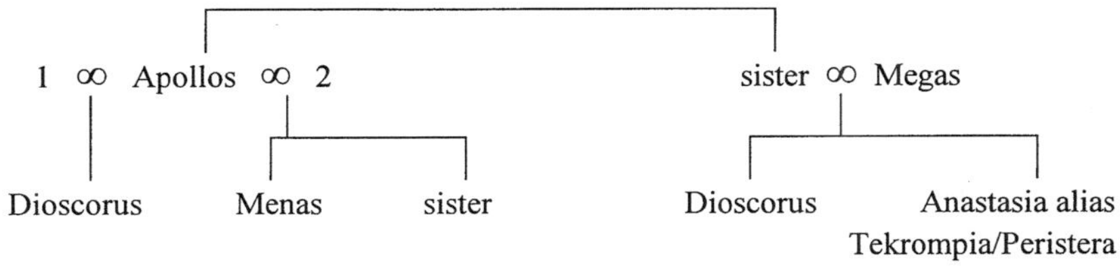

A WORLD FULL OF THE WORD

THE BIBLICAL LEARNING OF DIOSCORUS

Jitse H.F. Dijkstra

As we live, move and are in God, so let us search for God in all our studies, taste his goodness, touch his miracles, sigh to him with short and humble prayers, and therewith season everything, whether we read or write, or dedicate ourselves to contemplation, as if with sweet honey, as if with spiritual herbs and spices.[1]

These are the words that Gysbertus Voetius, the most prominent figure of the Dutch movement known as the *Nadere Reformatie*, spoke at the end of his inaugural lecture on the occasion of his appointment as professor of theology at Utrecht University on 21 August 1634. And indeed, Voetius's writings are full of biblical learning. In his period, the Bible was an instrument through which to express one's feelings, a feature it has in common with Late Antiquity.

Late Antiquity witnessed one of the most profound cultural changes in history: the Christianisation of the Roman Empire. From the reign of Constantine onwards, when Christianity became the favoured religion, the new faith spread over all segments of society. Although this process seems surprisingly efficient and successful at first sight, it by no means meant a rapid triumph over old norms and values. The Christianisation process was a long and complicated development of appropriation and assimilation, in which classical Antiquity was always looming large. Not surprisingly, within this cultural transformation, language changed in a similar way. Christian words, and combinations of words, were incorporated in all kinds of texts, from literature to documents. In Egypt, we can closely follow this tendency by studying the many thousands of papyri which have survived the depredations of nature and man. Under close scrutiny, it appears that Late Antique texts were not totally different from those of preceding periods. It was the meanings of words, and of combinations of words in particular, that had changed.

Professionally, Dioscorus (c. 520-585 AD) lived in a world full of words. As a lawyer composing documents, he was daily concerned with

[1] Voetius, *Oratio inauguralis*, pp. 51-52.

words. Moreover, from the fourth century onwards documents were written in an increasingly verbose and pompous style.[2] It was out of fashion to state things as they were. Documents were fuelled by rhetoric, in those days the classical rhetoric of Isocrates, with which men of letters were educated. These documents, of the sixth century onwards especially, strike us as overdone. Written in what we call the 'chancery style', they are wordy, formulaic, and emotional. Dioscorus wrote in just such a style. For example, the longest document of his archive, a testament, contains no less than 307 lines, forming a true *chef d'œuvre*.[3] If we ask ourselves the question, 'Why did Dioscorus not use fewer words?', we are approaching his archive from too modern a perspective.[4] For us, documents must state the facts without much ado. For Dioscorus, facts were clothed in words; and he was quite a master at using them.

In sixth-century Egypt, after two centuries of Christianisation, Christian idiom had become part of that vocabulary. Consequently, Dioscorus also had to be a master at using biblical words. This line of thought leads us to the central theme of this paper: biblical learning. How was Dioscorus educated? What do we find of biblical learning in his archive? And how was his biblical learning presented? After the Council of Chalcedon (AD 451), Egypt had become for the most part monophysite, so we may expect Dioscorus in his position at the court of the governor of the Thebaid to be associated with this current of Christianity.[5] But his position says nothing about Dioscorus's personal beliefs, and we shall make no attempt to resolve this question here.[6]

A quick glance at school texts from the fourth to the seventh century teaches us that classical education was still at the core of every school curriculum.[7] Homer's *Iliad*, Isocrates's speeches, and Aesop's fables are among the most cited works used for school practice. Under the influence of Marrou's *Histoire de l'éducation dans l'antiquité* (1948), it has been thought for a long time that education in Late Antiquity was mainly classical.[8] However, recent studies have demonstrated that biblical texts from

---

[2] Zilliacus, *Zur Abundanz*.

[3] *P.Cair.Masp.* II 67151 (15 November 570).

[4] See, for example, the comment on the mentioned testament (note 3) by its first editior, Maspero (*Papyrus grecs*): 'Ce papyrus est à ma connaissance le plus long, dans ce genre, qui ait encore été publié; étendue qui, malheureusement, est due surtout à la fatigante prolixité du style'.

[5] Frend, *The Rise*.

[6] Cf. Bowersock, *Hellenism in Late Antiquity*, pp. 63-69.

[7] I have consulted the appendix of Cribiore, *Writing*.

[8] Marrou, *Histoire de l'éducation*, pp. 425-426.

both the Old and New Testaments were used also at Egyptian schools.[9] The archive of Dioscorus confirms this practice and shows us that he was well educated in both the classical and the Judaeo-Christian tradition. The term for a man of letters was *scholasticus*, and it is exactly this title that Dioscorus has in one of the papyri.[10] It is hard to decide where he received his education. In the fifth century, the Thebaid flourished as one of the main centres of knowledge and learning in Egypt. It is therefore likely that Dioscorus received at least the first stages of his education here, and was subsequently trained as a lawyer in one of the famous schools of Alexandria.[11]

Traces of Dioscorus's education can still be found in his archive. It is the uniqueness of this archive that his poems, his documents and a small, personal library have survived. Hence we know that Dioscorus had some passages of the *Iliad,* including a commentary on them or *scholia minora*, a codex with comedies of Menander and one with fragments of other comic writers, a short life of Isocrates, and a passage from the *Anthologia Palatina* at his disposal. Additionally, several texts with conjugations, a Greek-Coptic glossary, a medical recipe, a text with incantations against demons and a meteorological table have been found in the archive.[12] These texts probably served a modest educational purpose, but they also reflect Dioscorus's own education, for they return frequently in his poetry.[13] In an iambic poem, he even praises Isocrates and Menander as his rhetorical models.[14] Thus, Dioscorus was educated according to the canons of his age.[15] Strikingly absent from his archive, however, are biblical texts. How could this be? Was Dioscorus's archive thrown away at some point after his death? Were the Old and New Testament texts kept behind for a local monastery like the one founded by Dioscorus's father, Apollus? We simply do not know.

Whether Dioscorus actually possessed biblical texts or not, biblical learning is strikingly, though not entirely, absent from his poetry. A simple solution for this problem is that Dioscorus was writing his poems in the

---

[9] Bagnall, *Egypt*, pp. 103-104; Cribiore, *Writing*; Bucking, 'Christian Educational Texts'; Browning, 'Education', p. 856; Cribiore, *Gymnastics of the Mind*, pp. 40, 106.

[10] *P.Cair.Masp.* I 67064.13. The term *scholasticus*, Greek σχολαστικός, denotes both a man of letters and a lawyer. Cf. Fournet, *Hellénisme*, p. 688, note 98; Claus, *'Ο σχολαστικός*; Keenan, 'The Names Flavius and Aurelius', p. 301 and Hickey, 'A Fragment of a Letter'.

[11] Cameron, Averil, *Literature and Society*, pp. 472-473.

[12] Fournet, *Hellénisme*, pp. 1-237.

[13] Fournet, *Hellénisme*, pp. 669-690.

[14] *P.Aphrod.Lit.* IV 4. See Van Minnen, 'Isocrates and Menander'.

[15] Fournet, *Hellénisme*, p. 672.

classical tradition. Above all, in his study of the Dioscorean poems, Jean-Luc Fournet has demonstrated the continuity of Hellenism in the sixth-century: 'Dans l'ensemble, la culture livresque de notre poète est sans originalité, fidèle reflet des canons littéraires de son époque'.[16] The great fifth-century poets of his region are known as the 'Wandering Poets', and of the latter Nonnus most directly influenced Dioscorus.[17] When we look at the two preserved works of Nonnus, we can draw interesting parallels with the Dioscorus archive. Nonnus's first work, the *Dionysiaca*, consisted of forty-eight books, in which Dionysus' conquest of India formed the main part (books 13-40).[18] There is nothing overtly Christian in this work. But then there is Nonnus's other work, written in a similar style: the *Paraphrase of the Gospel of Saint John*.[19] There is no reason for us to become trapped in discussions about whether Nonnus was or became a Christian, and neither shall we discuss here the same question for Dioscorus.[20] Apparently, in a Christianising world it was possible to write epic poems in a Hellenic style, and for other purposes to compose a paraphrase of a gospel.[21] That is why Dioscorus composed Hellenic-style poems, but wrote documents with references to Christian texts. This is not to say that the poems lack Christian thought. Some motifs in the poems are typically Christian.[22] One poem, though not written by Dioscorus himself, is dedicated to the Christian saint Menas, who had an important pilgrimage site in Alexandria.[23] But over all, the purpose of these poems was diplomatic: to praise a high official in an elaborate, Hellenic-style poem.

When we turn to the documents, the picture changes entirely. Amphilochios Papathomas has recently demonstrated that Dioscorus's documentary texts are deeply influenced by New Testament vocabulary.[24] Although such a study does not yet exist for the Old Testament, scholars have recognised some references in petitions as well.[25] Why do we find biblical learning in

---

[16] Fournet, *Hellénisme*, p. 683.
[17] Fournet, *Hellénisme*, pp. 678-680.
[18] Vian, *Nonnos de Panopolis*; Hopkinson, *Studies in the Dionysiaca of Nonnus*; Shorrock, *The Challenge of Epic*.
[19] Scheindler, *Nonni Panopolitani Paraphrasis S. Evangelii Ioannei*; Livrea, *Parafrasi del vangelio*; Hilhorst, 'The Cleansing of the Temple'.
[20] Livrea, *Parafrasi del vangelio*, pp. 19-35 for a full bibliography. Cf., most recently, Cameron, 'The Poet'.
[21] Bowersock, *Hellenism*, pp. 41-53.
[22] Fournet, *Hellénisme*, pp. 341-342, 682.
[23] *P.Aphrod.Lit.* IV 48.
[24] Papathomas, 'Zwischen juristischen Formeln'.
[25] Fournet, *Hellénisme*, p. 681, note 63, adding *P.Cair.Masp.* I 67002 ii 16, cf. Dt. 12.16, 24, 15.23.

the documents? We have noted that from the fourth century onwards documents became increasingly wordy, formulaic and emotional. Especially in documentary types in which these characteristics were meaningful, like testaments, letters and petitions, the dividing line between literature and document became blurred, in what is called by Fournet 'la "littérarisation" d'un genre'.[26] However, in no document was the art of rhetoric more appropriate than in petitions. Petitions needed to impress the addressees, mostly high officials, and themselves well educated men. Petitions needed to contrast the 'good guys' (the petitioners) with the 'bad guys'. In petitions, sick people were always 'half-dead'; in petitions, petitioners were always 'miserable' or 'humble slaves'.[27] It is in this type of document that rhetoric flourishes most in this period. Using biblical references was one of the options. No wonder, therefore, that the majority of references in Papathomas's article are to petitions.

The petitions of the Dioscorus archive, then, offer us the best opportunity to look at how Dioscorus works with biblical texts. Before embarking upon an analysis, however, we have to make some preliminary remarks about the appropriate way to approach biblical references in such texts. How do we know that an allusion is intentional? We only know this for sure if a verse, phrase or line is directly cited from the Bible, or a biblical name is mentioned. Unfortunately, until now in the Dioscorus archive only two direct quotations of the Old Testament have been discovered, and those consist only of two names referring to the Old Testament.[28] We have to figure out the rest ourselves, which implies in itself a well educated audience. But we should not over-estimate the author and his public. In a recent study, Clement Kuhn has argued that Dioscorus's poems have a deeper, mystical meaning. In his own words: 'the complete allegorical subtext was meant not to be fully comprehended by the government administrator but only by a more learned, sensitive, or spiritually "initiated" audience'.[29] However, Fournet's work has shown that the function of Dioscorus's poems was often diplomatic and that they probably sometimes even accompanied documents which were handed over.[30] That makes a 'subtext' superfluous. We also

---

[26] Fournet, 'Entre document et littérature'.

[27] I do not see the development perceived by Kovelman, 'From Logos to Myth', p. 139: 'Therefore, in late antiquity, typology ('moderate' – 'strong') was replaced by analogy (sometime – now) which better suited the contemporary historical perception of the world'.

[28] Direct citations: *P.Flor.* III 296.17-19 = Ps. 40 (41) 2-3 and *P.Cair.Masp.* I 67007.19 = Ps. 102 (103) 4 (personal communication with J.-L Fournet); names: *P.Cair.Masp.* I 67002 i 18 (Midianites) and *P.Cair.Masp.* I 67004.11 (Jeroboam I).

[29] Kuhn, *Channels of Imperishable Fire*, p. 159.

[30] Fournet, *Hellénisme*, pp. 685-686, 688.

have to be careful not to underestimate author and audience. To return to the petitions, Arkady Kovelman has tried to see influences from the Greek novel in a petition.[31] Although such influences are not unthinkable, the comparison is too vague to be convincing.[32] Therefore, if we wish to analyse a text with regard to biblical allusions, we have to consider: 1) whether the word or words are only used in the Bible; 2) whether the connotation of a word is meaningfully Christian; and 3) whether the context makes such an allusion probable. Only by carefully balancing these considerations can we pick out the meaningful allusions. And even then this interpretation may be doubtful.

The text under consideration here is *P.Cair.Masp.* I 67004. It belongs to a group of thirty-one petitions, all drafts, produced when Dioscorus was νομικός or notary in Antinoopolis, the capital of the Thebaid (c. 565-573). Two other petitions from the hand of Dioscorus are earlier (546/7) and request the help of the Empress Theodora.[33] They are less elaborate in style than the later ones, and biblical language is not so easily recognised. Apparently, Dioscorus needed rhetoric more at the Antinoopolite court than at home. From these petitions, *P.Cair.Masp.* I 67004 nicely illustrates how Dioscorus works with biblical texts. The twenty-one lines of the papyrus are comparatively well preserved, although on the right-hand side twenty-eight letters are missing and insects have made of its material a welcome meal. It belongs to a small group of petitions in which we still can make out the main line of thought. Moreover, concentrating on one petition has the advantage of allowing one to discuss passages in their context instead of presenting isolated cases from different petitions. In what follows, I first give a brief introduction to the papyrus and its contents. Then I show how Dioscorus stigmatises the 'bad guy' of the story with words that have a specific, negative connotation. Finally, I give three examples of Dioscorus's biblical learning, with an increasing degree of probability that these allusions are indeed deliberate.

The first editor of our petition, as of so many other papyri belonging to the Dioscorus archive, was Jean Maspero, who presented the text along with a very concise commentary in the first of three volumes containing Dioscorus papyri in the *Catalogue général* of the Egyptian Museum of Antiquities in Cairo.[34] Owing to its fragmentary state, this text has received less attention than it deserved. The present writer is preparing a re-edition of the

[31] Ach.Tat. 6.21, cf. *P.Cair.Masp.* I 67005.17-8.
[32] Kovelman, 'From Logos to Myth', p. 141.
[33] *P.Cair.Masp.* I 67019 and III 67283. See Gagos and Van Minnen, *Settling a Dispute*, pp. 10-15.
[34] Maspero, *Papyrus grecs*.

text with translation and commentary. The petition can be dated to 567 or 568, and is addressed to the highest official in the Thebaid, the *dux*. It concerns the town councillors of the city of Omboi in the southern Thebaid. They complain about a man who is preventing them from collecting taxes. Henceforth, I refer to him by his Dioscorean nickname, ὠμόφαγος, literally 'Eater of Raw Meat' (lines 10 and 14) or 'Cannibal'. As councillors, the petitioners are responsible for regional tax collection. So far so good. Tax collection is one of the themes to be expected in the documentary papyri from Egypt. The petitions have in common that they all describe a similar disruption of peace, which has to be corrected.[35] We will see, however, that Dioscorus elaborates on 'attending circumstances' to make the situation, and thus those around him, even more unbearable.[36] In our petition he does this by depicting this Cannibal as a lawless person.

Petitions of this period start with a formal address (l. 1) and the names of the petitioners (l. 2). The rest of the petition follows the structures of rhetoric. A new element in Late Antiquity is that petitions have an *exordium* in which the case is set forth (ll. 3-6). The *exordium* often depicts the addressee as the new Christ, who will redeem his subjects. In our petition this is done by describing the arrival of the *dux* as the arriving Christ (παρουσία, l. 4). The *narratio* is the freest part of the petition (ll. 6-17). It falls into two sections, in which Cannibal twice ravages the country and acts according to his own laws (αὐτον[ο]μί[α]ς τρόπῳ in l. 11 and αὐτονομῶν l. 14). In the first section, Dioscorus focuses our attention on how Cannibal persuades a group of desert people, called the Blemmyes, to return to their pagan worship and to plunder the countryside with him (ll. 6-11). In the second section, he again destroys the countryside, now when the councillors are in Antinoopolis for taxation business (ll. 12-17). With his gang, Cannibal even spends taxation money on his own behalf. The *dux* should make no mistake: this malefactor needs to be stopped. In the last part of the petition (the *conclusio*, ll. 18-21), the petitioners ask this favour of the *dux*. It should

[35] Dioscorus himself well summarises this basic thought of his petitions in one of the poems (*P.Aphrod.Lit.* IV 10.1-4):
Θήβη, πασᾶ χόρευσον, εἰρήνην δέχου·
οὐ γὰρ θεωρήσῃς κακουργικὴν ἔτι,
οὐ βαρβάρων δέος, φιλοπραγμόνων κρίσιν·
πάντη γὰρ Εἰρήνη θεόπνευστος ῥέει.
'Thebes, dance altogether, receive peace:
for no longer will you see malicious deeds,
not fear of barbarians, nor arbitrariness of the greedy;
for everywhere flows a god-inspired Peace.'

[36] Another good example of such a petition is *P.Cair.Masp.* I 67002.

come as no surprise that the *narratio* is the place to 'set forth the case', and it is on this part of the petition that we now focus.

In the whole *narratio*, words abound that have a negative moral connotation. Dioscorus uses classical words, but also words that have a Christian connotation of someone who lives beyond the Law of God, and they remind us of the New Testament *Lasterkatalogen*.[37] For example, some words in the petition evoke 1 Timothy 1.3-11, which is exactly about people who diverge from God's law and teach their own (νομοδιδάσκαλοι, 1 Tim. 1.7), like νοσέω 'to be sick' (l. 6), corresponding to 1 Tim. 6.4,[38] and βέβηλος 'unholy' (l. 16), corresponding to 1 Tim. 1.9. Moreover, in the Bible νοσέω only has the sense of an abnormal condition of the inner self in this context.[39] Other words having a strongly negative connotation in the Bible are μωρία 'foolishness' (l. 6),[40] ἀκαταστασία 'instability' (l. 6),[41] ἀταξία 'disorder' (ll. 6, 15),[42] μάταιος 'fool' (l. 7),[43] παρανομία 'transgression of law' (l. 12),[44] and ἐπιτιμία 'penalty' (l. 12).[45] Furthermore, with a typically biblical term Cannibal σκάνδαλα ποιέω (l. 9) 'laid traps' which is the equivalent of the New Testament verb σκανδαλίζω 'to bring somebody down'.[46] Finally, the verb συνθρησκεύω (l. 16) 'to worship together' has the same context as θρησκεύω in Wisdom 11.15 and 14.16, which are also passages about the worship of idols. True, these words do appear in classical texts, but the well educated audience of Dioscorus would have surely recognised the Christian connotation.

Allusions to biblical passages can also be found in ll. 7-9, the text of which it is worth quoting in full:

> Οὐκ ὤκνησεν γὰρ ὁ τοιοῦτος μάταιος καὶ ọλισθηρὸς καὶ [± 28] πονηρὸν ἀπεργά[σασθαι] ἑαυτῷ βίον τε καὶ ἐργαστήριον, καὶ τὸ μεμυημένον [..] χριστιανικὸν ἀθετῆσαι σέβας καὶ θρῆσκος, καὶ δαίμοσι καὶ ξοάνοις ἀφιερῶσαι <u>σηκούς</u>, ἀντιπολέμιον ἑαυτὸν δ[εικνύμενος ± 18] σκάνδαλα ποιησ[άμενο]ς, ἐν τῷ αὐτὸν <u>τὰ ἱερὰ</u> τοῖς βαρβάροις ἤτοι Βλέμυσι

---

[37] E.g. Fitzgerald, 'Virtue/Vice Lists'.

[38] νοσῶν περὶ ζητήσεις καὶ λογομαγίας 'he has a morbid craving for controversy and disputes about words' (RSṼ)

[39] Kittel *et al.*, *Theologisches Wörterbuch*, *s.v.* Cf. οὐκ ὡς ἔτυχεν 'in an abnormal way' (l. 5). See for the following words also the entries in the *ThWNT.*

[40] Sir. 20.31; 1 Cor. 1.18, 21, 23; 2.14; 3.19

[41] Prov. 26.28; To. 4.13; Luke 21.9; 1 Cor. 14.33; 2 Cor. 12.20; James 3.16.

[42] 1 Thess. 5.14, 3.6, 11; 2 Thess. 3.7.

[43] Act. 14.15; 1 Cor. 3.20, 15.17; Tit. 3.9; James 1.26; 1 Pet. 1.18.

[44] Ps. 36 (37) 7; Prov. 5.22, 10.26, 26.7; 4 Macc. 2.11, 4.19, 5.13, 9.3; 2 Pet. 2.16.

[45] Wis. 3.10; 2 Cor. 2.6.

[46] E.g. Matt. 18.7; Luke 17.1.

δ[ι]ακαινίσασθαι, μεθ' ὃ καίπερ ἐκεῖνοι τὸν ἀίδιον ἐπιγινώσκειν Θεὸν ὑπεστρέφοντο τῶν ἐξουδενώσεως εἰδώ[λων ± 25]

For that fool, slippery and ... person ... did not hesitate ... to make his own life and workplace evil, to set aside the mysterious Christian worship,[47] and to dedicate shrines with demons and wooden statues, (thereby) showing himself an adversary ... and he laid traps by renewing the sanctuaries for the barbarians, that is, the Blemmyes, after which he, although they were turning away from the despicable idols towards the knowledge of the eternal God ...

This passage is remarkable, for it tells us that Cannibal brought a group of Blemmyes from the Christian faith by helping them recover an old shrine of theirs. Two biblical echoes can be heard here. Firstly, the combination σηκός - ἱερόν is only found in 2 Maccabees 14.33. It is the story at the end of the second book where the Ptolemaeic general Nicanor goes up to the temple of Jerusalem to imprison Judas Maccabaeus. When the priests swear that he is not there, Nicanor threatens:

'Εὰν μὴ δέσμιόν μοι τὸν Ιουδαν παραδῶτε, τόνδε τὸν τοῦ θεοῦ σηκὸν εἰς πεδίον ποιήσω καὶ τὸ θυσιαστήριον κατασκάψω καὶ ἱερὸν ἐνταῦθα τῷ Διονύσῳ ἐπιφανὲς ἀναστήσω.

If you do not hand Judas over to me as a prisoner, I will level this precinct of God to the ground and tear down the altar, and I will build here a splendid temple to Dionysus (RSV).

Of course, threatening to raze to the ground the Temple of Jerusalem spells trouble. In the subsequent episodes, Judas beats Nicanor in battle and carries his head and arm with him to Jerusalem as trophies: 'and he cut out the tongue of the ungodly Nicanor and said that he would give it peacemeal to the birds and hang up these rewards of his folly opposite the sanctuary' (2 Macc. 15.33). Although the combination σῆκος - ἱερόν is plural in our text, all we can say is that the combination never occurs elsewhere and may have rung a bell in the audience.

A more prominent allusion in this passage is to 2 Peter 2. In this chapter, Peter warns against false prophets (ψευδοπροφῆται, 2 Pet. 2.1). The *portée* is strikingly similar to the passage in our petition, for in both texts 'false prophets' bring other people from the way of truth and lead them to lawlessness. The texts have a series of words in common: ἀνόμοις (2 Pet.

[47] μεμυημένον 'initiated into the mysteries (baptised)', hence 'mysterious'; σέβας καὶ θρῆσκος (read θρησκεία) 'worship': *hendiadys*.

2.8) ~ μηδεμίαν ἔκπληξιν τῶν νόμων καρτούμενος (l. 7), αὐθάδεις (2 Pet. 2.10) ~ [αὐθ]αδέως (l. 17), ἀδικίας (2 Pet. 2.13, 15) ~ ἀδικημάτων (l. 4), παρανομίας (2 Pet. 2.16) ~ π[αραν]ομίαν (l. 12) and ματαιότητος (2 Pet. 2.18) ~ μάταιος (l. 7). After describing where lawlessness can lead, Peter returns to the false prophets and how they lead converted people astray. Lapsing after having been converted, is even worse than being a non-convert in the first place, says Peter (2 Pet. 2.21):

> κρεῖττον γὰρ ἦν αὐτοῖς μὴ ἐπεγνωκέναι τὴν ὁδὸν τῆς δικαιοσύνης ἢ <u>ἐπιγνοῦσιν ὑποστρέψαι</u> ἐκ τῆς παραδοθείσης αὐτοῖς ἁγίας ἐντολῆς.
>
> For it would have been better for them never to have known the way of righteousness than after knowing it to turn back from the holy commandment delivered to them.

Although it is hard to compare both sentences because of the lacuna immediately following the subordinated καίπερ clause in our text, the verbs ἐπιγινώσκειν and ὑπεστρέφοντο strongly evoke 2 Pet. 2.21. However, the meaning is the other way round. 2 Pet. 2.21 is about people turning away from the true faith after having been converted. The petition is about people turning away from idolatry towards the Christian faith, that is, they were about to be converted. The flow of thought would then be that Cannibal renewed the sanctuaries for the Blemmyes, after which he brought them from the faith, which is even worse, because they were just converting to Christianity. To sum up, the καίπερ sentence with its allusions to 2 Pet. does not, like the apostle, refer to the turning away from the true faith, but to the conversion to Christianity which the Blemmyes had experienced apparently not long ago. In this way, Dioscorus varies on 2 Pet., though the passages run parallel in that Cannibal leads the recently converted from the way of truth.

In the next two lines of our petition (10-11), the consequence of the false prophecy of Cannibal seems to be that he lured the gang of Blemmyes into plundering the countryside with him. In the climax of the first section, Dioscorus states that Cannibal acted for the first time 'according to his own laws': αὐτον [ο]μί[α]ςτρόπῳ καθ ['Ι]ερωβοὰμ τὰς δαμάλε[ις ± 26].

Dioscorus is here comparing Cannibal to Jeroboam, who was the first king of Israel after the disruption of the kingdom of Solomon.[48] Afraid of losing his kingdom if his people should continue to worship God in the

[48] 1 Kg. 12.19: ἠθέτησεν 'Ισραὴλ εἰς τὸν οἶκον Δαυίδ 'So Israel had been in rebellion against the house of David'. Dioscorus uses the same verb to indicate Cannibal's "disruption" of Christian worship (8). Cf. 1 Chr. 2.17; 1 Kg. 8.50; Heb. 10.28.

temple of Jerusalem, he set up two golden heifers, one in Bethel and one in Dan (ἐποιήσεν δύο δαμάλεις χρυσᾶς, 1 Kings 12.28). This became the reason for sin (καὶ ἐγένετο ὁ λόγος οὗτος εἰς ἁμαρτίαν, 1 Kg. 12.30). He erected temples, ordained priests and festivals, and made sacrifices to the heifers. One day, a man of God came to prophesy against the altars, but Jeroboam did not desist from his godless deeds (1 Kg. 13.33-34):

> Καὶ μετὰ τὸ ῥῆμα τοῦτο οὐκ ἐπέστρεψεν καὶ ἐποιήσεν ἐκ μέρους τοῦ λαοῦ ἱερεῖς εἰς τὰ ὑψηλά. Καὶ ἐγένετο τὸ ῥῆμα τοῦτο εἰς ἁμαρτίαν τῷ οἴκῳ Ἰεροβοὰμ καὶ εἰς ὄλεθρον καὶ εἰς ἀφανισμὸν ἀπὸ προσώπου τῆς γῆς.

> After this thing, Jeroboam did not turn away from his evil way, but made priests for the high places again from among all the people; any who would, he consecrated to be priest of the high places. And this thing became sin to the house of Jeroboam, so as to cut it off and destroy it from the face of the earth.

The comparison between Jeroboam and Cannibal holds good, for both the Israelite Jeroboam and the Christian Cannibal sin by bringing believers from the faith. The message is clear: Cannibal's godless and lawless behaviour will eventually lead to his destruction.[49]

We have seen that in the first section of the *narratio*, Dioscorus to the fullest extent possible exploits biblical allusions to depict Cannibal as an autonomous, and consequently dangerous, person for what this petition is all about: collecting taxes. Two themes that are elaborated in the second section, are his conspiracy with the barbarian Blemmyes, and his godlessness. Both themes are summarised in an expression that moreover well illustrates the changing idiom of Late Antiquity. Cannibal is described as a μιξοβάρβαρο[ς κ]αὶ μιξ[έλ]λην (l. 14). The antithesis Ἕλληνες - βάρβαροι was one of the most well known word couples in classical literature, which, when used together, could mean 'everybody, not even barbarians excluded'.[50] In Late Antiquity, however, the word Ἕλλην had changed meaning and was used as an expression for a heathen or pagan.[51] Similarly,

---

[49] The story of Jeroboam has a rich afterlife in patristic texts: Iren. *Haer.* 4.62.2; Tert. *Jud.* 1.7; Or.*Ep.* 2 3.66.13, 15, 22; *Fr. in Jer.* 221.12; 226.14; *Hom. in Ezech.* 9.1; *Hom. in Gen.* 15.7; Bas.*Enar.Es.* 445 C 14, 448 B 1, 5, 515 B 8; Greg.Naz. *Or.* 5.3 (in which Jeroboam and Ahab are called τοὺς παρανομωτάτους, cf. παρανομίαν in l. 12 of our text), 6.36.5; Amph.*Haer.* 15.

[50] Cf. D. 18.202: τίς γὰρ οὐκ οἶδεν Ἑλλήνων, τίς δὲ βαρβάρων; 'For who does not know it from the Greeks, who from the barbarians?' Cf. Kemmer, *Die polare Ausdrucksweise*, pp. 88-90; Zilliacus, *Zur Abundanz,* pp. 32-33.

[51] Cf. the fifth-century Wandering Poet Olympiodorus of Thebes, who is called Ἕλλην τὴν θρησκείαν 'practising pagan' (*test.*, Blockley).

people living on the fringes of the known world, barbarians, were associated with paganism.[52] Dioscorus thus describes Cannibal as a 'half-barbarian and half-heathen', in the sense that he mingles with both. The two negative compositions even enhance each other in derogatory meaning, as if to say: How low can you go?

In conclusion, we have seen how Dioscorus uses his biblical learning to reinforce his petitions. Other petitions from the fourth until seventh centuries are less clear in unveiling biblical learning.[53] Was Dioscorus the only notary to exploit his biblical learning in the petitions? This would seem improbable, as there were certainly other notaries in Egypt and Antinoopolis with an education equal to, or better than, that of Dioscorus. In a fourth-century petition to a tribune in Oxyrhynchus, a passage from Jeremiah is referred to.[54] There is, then, no reason to make Dioscorus a special case.[55] What is so special about Dioscorus is that his petitions have survived, be it in the form of drafts. Let us therefore judge Dioscorus on his merits and approach his documents as documents of their time.[56]

---

[52] E.g., Rives, 'Human Sacrifice'.

[53] With the exception of the petitions in Schwartz (*Acta Conciliorum*), but I have limited myself here to petitions in an administrative context.

[54] *P.Oxy.* L 3581.13 ἐρήμωσιν εἰργάσατο κατὰ τοῦ οἴκου μου 'worked destruction for my house' ~ Jer. 22.5 εἰς ἐρήμωσιν ἔσται ὁ οἶκος οὗτος. This allusion is reinforced by 3581.12 ὀρφανίας ~ Jer. 22.4 ὀρφανόν. The allusions which Kovelman, 'Logos to Myth', pp. 150-152 mentions regarding *P.Oxy.* I 131 (fifth-sixth centuries) and XXVII 2479 (sixth century) are possible but unconvincing.

[55] Cf. Papathomas, 'Zwischen Juristischen Formeln', p. 497: 'Daβ nur Dioskoros sich traut, in seinen Verträgen die Formulierungen des griechischen Urkundenwesens zu variieren, und seine Petitionen in dem Maβ mit literarischen Reminiszenzen zu bereichern, liegt an der starken Persönlichkeit des durch sein literarisches Talent sprachlich souveränen und durch seinen wirtschaftlichen Wohlstand und seine politische Erfahrung sozial anerkannten und selbstsicheren Dioskoros'.

[56] I thank Prof. J.N. Bremmer, Prof. M.A. Harder, Dr. A. Hilhorst and Prof. P. van Minnen for reading an earlier draft of this paper.

LITERARY CULTURE IN SIXTH-CENTURY EGYPT

Teresa Morgan

Many years ago, Late Antique Egypt was regarded as an area in decline, still productive of grain and other goods, but socially and culturally degraded. Towns were shrinking and cultivated areas returning to desert; less and less Greek literature was being read and the poetry and other literary products of the time were best forgotten.

More recently, the emphasis has been on change rather than decline. On this view, Late Antique Egypt had a vigorous culture all its own, drawing on Egyptian, Greek and Christian materials. It has been emphasised that the decline of towns and abandonment of cultivated desert areas was no greater than in other areas of the empire at the time, perhaps less. The people of the provinces, moreover, played a full part in the wider Roman empire, travelling to Rome and Constantinople and elsewhere, meeting emperors and holding high office.[1]

Such different views are to some extent a matter of perspective: the generation which saw decline in Egypt saw it everywhere in the Roman world, while, these days, the fashion is to emphasise continuity and success. To some extent, they depend on the evidence one uses: the micro-history of Aphrodite, for instance, presents us with a picture of one flourishing sixth-century town with dozens of churches, monasteries and civic buildings, a large and complex economy, and well-connected inhabitants.[2]

In this paper I focus on one aspect of sixth-century life, the state of literary culture. Here, it is true that Greek culture, for instance, is less obvious in the records than it was. On the other hand, in the fifth and sixth centuries we have records of a remarkable clutch of writers and scholars who made their mark on the wider empire, though few of their works survive. I have nothing new to add to studies of politically and culturally influential individuals such as Cyrus, Olympiodorus or Nonnus, or even Dioscorus himself. Instead I want to examine literary culture as a whole across Egypt, by a method which until recently has not been easily available to scholars, but is now both viable and productive of very interesting results.

---

[1] Bowersock, *Hellenism*, chapt. 5; Cameron, 'Wandering Poets'; *idem*, *Procopius*, pp. 19-25; *idem*, *The Mediterranean World*, pp. 131-138.

[2] Sketched by MacCoull, *Dioscorus*, pp. 1-15.

What follows is a survey of literary papyri in Greek and Coptic,[3] extracted from the *Leuven Database of Ancient Books* (= *LDAB*), an incomparable resource published on CD-Rom in 1998, and updated in 2001, by Prof. Willy Clarysse.[4] I examine which fragments survive of which authors, from which places and in what numbers. On the basis of this information I venture some generalisations about the state of literary culture in sixth-century Egypt and how it has developed from previous centuries, if at all. I consider whether people are reading the same authors in the days of Dioscorus as in previous centuries, or different ones, or the same ones in different proportions. I try to see whether literary culture moves about the country or whether the same towns and cities are central to cultural life in this century as earlier. I look specifically at educational finds, to see whether we can speculate about levels of literacy in general at this period, or on who is getting an education. And I look at a selection of texts in more detail to see what we can deduce from their appearance about their use.

My picture inevitably depends partly on Egypt-wide statistics, but partly on the products of a small number of larger archives which contain literary material. I therefore consider whether such archives have distorted our view of culture in this period. The archive of Dioscorus at Aphrodite is only one of several finds. Others come from the White Monastery at Panopolis, Mons Epiphanius near Diospolis (Thebes) and St. Catharine's Monastery at Sinai, and there are other clusters of texts from probably libraries or archives, at Tura, Genizah and Panopolis town.

I do not try to distinguish between Greek and Coptic culture. This is partly because it is now widely accepted that the two were much intermingled, especially in the culture of the wealthier and more educated, and partly because the majority of Coptic texts in the LDAB are bilingual or parallel versions of the same text – such as schooltexts. The most likely inference is that they come from a common cultural milieu. There are around fifty texts in Coptic in the LDAB, the majority (twenty-seven) biblical. There is a handful of translations of Greek literature – Homer, Menander *sententiae*, Chariton – in bilingual texts, and a dozen or so mixed other writings – magic, prayers, hymns, writing exercises, and so on. Their find-spots, where known, put them comfortably in line with Greek finds, except that more are found on average at monasteries: nine from Diospolis, three

---

[3] I.e. not documents, but including 'sub-literary' material such as schooltexts, prayers and hymns.

[4] At the time of writing I have not had access to the updated version, in which marginal corrections may have been made to the list of texts. I have, however, removed two fragments of Gaius from the list as wrongly attributed to the sixth century. The earlier version can be accessed on the web at http://ldab.arts.kleuven.ac.be/, for which reason I have not listed all 1010 texts here.

each from the Fayum, Hermoupolis and Panopolis, two from Deir-el-Gizaz and one each from Antinoopolis and Aphrodite.

Several general caveats attend this exercise. Many texts cannot be dated precisely by century, so I have included texts dated to the fifth/sixth and sixth/seventh centuries, with the result that my picture is of what one might call the 'long sixth century'. For this reason, my figures overall are likely to be too high. But since this applies to any century one chooses to discuss, the sixth century figures are not distorted relative to other centuries, and it is relative levels of culture that I am discussing here. Including all possible texts, the overall sample is just over 1000 texts. This is not large, but it is large enough to be worth examining and to let us draw some conclusions. The sample also, of course, reflects excavation patterns, which add a further significant element of distortion. Again, however, in many cases this problem is minimised by the fact that I am comparing findspots with their record in earlier centuries, and by the fact that most of what are thought to be the most important cultural centres of the late Roman period have been excavated. That said, there is doubtless more unexcavated literary material everywhere, which might change the picture I draw below. But for the time being, we do what we can with what we have, which is by no means negligible.

From the 'long sixth century', 1010 literary texts of all kinds survive. This compares with survivals from other 'long' centuries, as follows: from the first century AD, 829 texts survive; from the second, 2229; from the third, 1803; from the fourth, 975; from the fifth, 963; from the seventh, 624. The first century AD is low in every kind of survival, and in general, as we go through time, more papyri survive, so the falling trend in numbers of literary texts as we go through the Roman period masks a much sharper fall in real terms. This observation also applies to Christian texts: as we go through time, fewer and fewer people seem to read anything.

Having said that, the fall from the fourth to the sixth century is not steep: overall, reading is almost holding its own, against a background of periodic social, political and economic stress. There even appears to be a small rise in the sixth century.

This is also true region by region. If we take places where substantial numbers of literary papyri are found, there do not seem to be places where population holds up but literary culture declines sharply, or vice versa. The table below spells it out:

Date	C1	C2	C3	C4	C5	C6	C7
Place							
Fayum	123	300	195	83	95	104	79
Herakleopolite	4	–	5	5	16	15	10
Oxyrhynchus	256	970	764	197	138	69	9
Antinoopolis	–	14	50	54	51	71	16
Hermoupolis	13	49	63	74	93	76	29
Aphroditop.	0	3	8	4	2	40	–
Panopolis	1	2	12	20	18	12	15
Diospolis	5	13	12	29	19	70	74
Apollonop.	3	2	4	3	2	2	1
Sinai	–	–	–	7	6	15	10
Total	829	2229	1803	978	963	1010	624

So, for instance, we find the number of texts found at Oxyrhynchus dropping from the fourth century onwards, perhaps anticipating its gradual decline, while the number of texts at Hermoupolis Magna grows slightly, as it becomes one of the centres of late antique population and culture. Parts of the Fayum are holding their ground despite some depopulation, while texts at the monastery sites of Diospolis and Sinai grow steeply. Aphrodite, which shows the sharpest rise of all, is the findspot of the Dioscurus archive; without that, Aphrodite would show little change over the last four centuries of Roman rule. I shall return to the significance of the archive for our reading of cultural trends.

There are various ways in which we might measure literary culture. We can isolate educational material: evidence for people learning to read and write, reading some literature, and so on. We can look at survivals of literary texts in general, which we might break down into canonical authors (for which there is the most evidence throughout Roman Egypt), contemporary and more recherché authors, and authors used for particular purposes, such as religion. The reading of classical, especially canonical authors, is, in the Roman world in general, a strong index of literary culture.[5]

A range of authors survives from the sixth century. I have counted fragments from thirty-six identified classical or later pagan authors (including medical and legal writings), as well as many unidentified pieces. The great majority of those identified were from canonical or very famous authors, and they included Latin fragments of Cicero, Juvenal, Lucan and Vergil, as well as Latin jurists. There were identifiable fragments of eighteen Christian authors in addition to dozens of anonymous liturgic frag-

[5] Morgan, *Literate Education*, pp. 50-100.

ments, hymns and prayers, and at least two Jewish authors, Philo and Josephus.[6]

On the other hand, this range does not compare with the range of authors surviving overall, which runs into hundreds, and it focuses heavily on well-known authors (that is, authors of whom a large number of texts survive overall, or who are canonical), with a few eccentricities perhaps representing individual interests.[7] The lists below give, on the left, canonical authors with the number of their texts surviving, sixth-century numbers in brackets and overall survivals after them, and on the right, the same arrangement for other pagan authors:

Aeschines (1) *overall 52*
Apollonius Rhodius (2) *49*
Aristophanes (10) *50*
Callimachus (3) *71*
Cicero (1) *10*
Demosthenes (3) *161*
Euripides (14) *176*
Hesiod (2) *126*
Homer (c.58) *1119*
Isocrates (3) *84*
Juvenal (1) *1*
Menander (6) *152*
Philemon (1) *7*
Pindar (2) *53*
Plato (2) *90*
Sappho (1) *25*
Sophocles (4) *38*
Theocritus (4) *26*
Thucydides (3) *89*
Vergil (5) *5*

Aesop/life of Aesop (9)
Palatine Anthology (1)
Aristides (1)
Chariton (1)
Dioscurides (1)
Galen (1)
Heliodorus (1)
Hippocrates (4)
ps-Menander gnomologies (4)
Justinian (6)
Lucan (1)
Palladius (1)
Romanus Melodus (3)
Strabo (1)
Ulpian (2)[8]

We can see that most reading is of well-known authors, though some, like Sappho, Theocritus and Philemon are not otherwise well represented in pa-

6 Both authors read by Christians (Philo almost exclusively so), so they could come from Jewish or Christian readerships.

7 This does not help us to assess the social strata of sixth-century readers, as the most 'popular' authors were canonical authors such as Homer and the Bible, equally read by the highly educated, and what we might think of as 'popular' literature, such as novels, was certainly also, and perhaps only, read by elites (Cameron, *The Mediterranean World*, pp. 136-141).

8 A further 138 fragments are unattributed.

pyri, so are only doubtfully described as 'popular'. By contrast, texts appear of a wide range of otherwise little-known Christian authors, and of almost every book of the Bible.

One of the most interesting questions we can ask of this material is how survivals translate into reading communities. We have begun to see what was read. Where was it read? Do nearly all these texts of pagan authors, for instance, cluster in a handful of cultured cities? Is the reading of the Bible (or at least its presence) significantly more widespread? Are most texts represented in more than one place – do all reading or text-owning communities read the same things? Or do different communities read different authors?

The canonical classical author of whom most texts survive, predictably, is Homer. Homer also has a wide geographical range. In the long sixth century he appears (from north to south) in the Fayum (twice) Oxyrhynchus (five times), Antinoopolis (three times), Hermoupolis (nineteen times), Aphrodite (once), Dendera (once) and Diospolis (four times), as well as outside Egypt. The only important findspots not to be represented are the Herakleopolite and the monasteries of Panopolis and Sinai. In the latter, we should not be surprised to find Homer, but we need not be over-surprised not to. Outside some monasteries, however, we find what we should expect: that even at this late date, Homer is a constant among Greek-literature-reading communities, an author that educated people can still assume they have in common.

The next most copied texts at this date are Euripides and Aristophanes. Euripides, after Homer, is the most widely read Greek author. Representatives of his 176 identified texts appear at every spot where literature has been found, at some point, and somewhere in every century. There are fourteen texts dated to the 'long sixth century': a respectable proportion; clearly he is holding his popularity. Their known findspots, though, are not quite as widespread as one might expect. One text comes from Arsinoe in the Fayum; none from Oxyrhynchus; four from Antinoopolis; six from Hermoupolis and one, possibly, from Sinai. It is likely that the absence of finds from Oxyrhynchus is an insignificant accident, since much reading activity is visible there in the sixth century. Even if we notionally fill in Oxyrhynchus, however, Euripides does not appear, at this date, to be penetrating very far south. This may simply be the accident of survival, but it is worth noting for comparison with other authors.

Finds of Aristophanes follow much the same pattern. One comes from Arsinoe, two from Oxyrhynchus, two from Antinoopolis and three from Hermoupolis. Again, we see little southward penetration, though there is a respectable representation of texts from the last two middle-Egyptian cities, which became major sites of Late Antique culture.

Euripides, Aristophanes, and perhaps Sophocles and Theocritus seem to be holding their own best. Homer seems to be doing a little less well – perhaps as the most popular author he is losing ground to the Bible. Most other authors are in sharp decline. A sample of the authors of whom fewer texts survive, but of which the findspots are known, is statistically less useful, but still interesting. The findspots of Apollonius Rhodius, Hesiod, Philemon, Sappho and Strabo are regrettably unknown. Aeschines, Cicero, Demosthenes, Isocrates, Pindar and Vergil survive only, where known, at Oxyrhynchus. Aesop and Lucan, where known, survive only in the Fayum. Plato and Juvenal survive only in Antinoopolis. Thucydides survives in the Fayum and at Oxyrhynchus. Callimachus survives in Oxyrhynchus and Hermoupolis. Menander survives from Oxyrhynchus, Hermoupolis and Diospolis; Sophocles from Oxyrhynchus, Hermoupolis and Antinoopolis. Theocritus, of this collection, has the widest scope, surviving from the Fayum, Oxyrhynchus, Hermoupolis and Antinoopolis.

We have to be cautious when drawing conclusions from such slender data, but a pattern emerges here which is of interest. Of the fifteen authors of whom any findspots are known, two-thirds survive in only one place. In more than half those cases, the place is Oxyrhynchus. More than two-thirds overall (eleven) survive at Oxyrhynchus. Four survive from Hermoupolis and three from the Fayum. All the texts which survive in more than one place, do so in more-or-less contiguous areas. We do not find any text surviving, say, in the Fayum and Dendera, or Oxyrhynchus and Apollonopolis.

At Oxyrhynchus, it is clear that classical culture is still relatively vigorous at this time, and Oxyrhynchus yields the widest range of authors and the least widely-read. As we move south, the authors surviving are more mainstream: Homer, Euripides, Menander, Sophocles, Aristophanes, Callimachus, Theocritus. In the north, surviving authors include more rarified interests, such as Lucan.

There seem to me several possible models for the changing pattern of classical reading in Egypt. Did cultural communications degrade, leaving islands of readers diverging in their interests? Did cultural communications continue good but the repertoire of authors shrink? Did reading shrink round the edges of the country, so to speak, leaving a relatively full cultural picture in the middle, but decreasing as one went north or south?

On the basis of the data examined so far, I suggest a combination of all three. Clearly, reading shrinks around the edges of the provinces, and the repertoire of authors also shrinks. Some authors, such as Homer, continued to some degree to define cultural orthodoxy all over the country. Homer is the only author found in many southern towns – Aphrodite, Dendera – and almost the only one at Diospolis. By the sixth century, however, few if any other authors, not even Euripides and Menander, seem to have such power to define Greekness that they are universally read. There also seems to be some breakdown of consensus about what to read. It is notable how few texts are found, for instance, in both Oxyrhynchus and the Fayum at this date, though reading habits in parts of the Fayum clearly continued to be sophisticated. This suggests that communications between these two centres, for cultural purposes, may not have been as extensive as they had been in earlier centuries. On the other hand, there is a large overlap between Oxyrhynchus, Antinoopolis and Hermoupolis, suggesting that in the middle part of the country, communications may have been better.

A missing but crucial term in this picture must be Alexandria. To some extent, we can postulate that Alexandria explains the differences in reading habits and communications between different places. In the Hellenistic and earliest Roman period, many wealthy Alexandrians had country estates in the Fayum, which therefore came into contact with all the latest cultural fashions. From the second century AD, Alexandrians began to own more estates in the area of Oxyrhynchus, and later still around Hermoupolis and further south. It seems likely that these places then took over as the leaders of culture, with the added factor that Oxyrhynchus, in particular, became a major cultural centre in its own right. The Fayum, at this later time, saw fewer wealthy Alexandrians; it did not abandon Greek culture, but it ceased to be cutting-edge. In other words, communications were patchy, better between some areas than others. Where communications with Alexandria were best, communication between upstate cities was also better, and these cities retained more reading habits in common. In a further development in Later Antiquity, wealthy Alexandrians and Constantinopolitans who owned land in Upper Egypt came to visit it more and more rarely, so the overall contact of these towns with the elite lessened.[9] (One wonders how significant it was for the fact that Dioscorus stayed largely in Aphrodite, that one eminent family, the Apions of Egypt, Constantinople and elsewhere, is known to have spent time in Egypt, and in his own area, the Thebaid.)[10]

How does the distribution of biblical and other Christian texts compare to that of classical authors? Do reading habits exist in tandem, or are some places more heavily classical, and others more heavily Christian? In the case of Christian authors (not the Bible, hymns, prayers etc.), we are hampered by the fact that the great majority have no provenance. But where they have, we can see, first, that the places where more Christian authors are read, are also those where more classical texts are read, and second, that

---

[9] Hardy, *The Large Estates*.

[10] Flavius Apion II was *Dux* from 548; his father had been Prefect of Egypt and 'leading citizen' of Herakleopolite and Oxyrhynchus.

on the whole, different Christian authors appear in different places. For instance, Origen and Didymus the Blind are found at Tura, near Cairo. In the Fayum we find fragments of John Chrysostom, Gregory of Nyssa, Cyril of Alexandria, Abbot Isaias and Eusebius of Caesarea. At Oxyrhynchus we find Basilius Seleuceius, Hippolytus of Rome, Theophilus of Antioch and John Chrysostom again. Basilius of Caesarea appears at Antinoopolis. The only other identified Christian authors are found in monasteries. All this suggests that reading major Christian authors is rather a specialised literary pursuit, which flourishes where literary culture in general is strongest, but about which there is not widespread consultation or communication. This may be partly because one of the two largest centres is in the Fayum, which, we have seen already, seems to have been short of cultural communication with Alexandria or the more southerly towns and cities at this period.

Fragments of the Bible tell a different story.[11] By far the most popular book of the Bible is Psalms, with ninety-six texts surviving from the sixth century. They come from all over Egypt, from Tura and Genizah near Cairo, down through the Fayum, the Herakleopolite, Oxyrhynchus, Antinoopolis, Hermoupolis, Panopolis, Diospolis, the Nagada region and Wadi Natrun. Other relatively popular Old Testament books are also somewhat dispersed: the nine books of Daniel through the Fayum, Herakleopolite and Diospolis, the nine books of Genesis through Tura, the Fayum and Sinai, the six books of Kings from Genizah to Hermoupolis and the Wadi Natrun. Here, however, as with classical authors, fragments from south of Hermoupolis, outside monasteries, are relatively rare. The monasteries of Panopolis and Diospolis make the distribution of texts look wide, but fragments from towns in the south of the country are fewer than half a dozen.

Texts of the New Testament show a similar scatter. The table below shows findspots of the most commonly found texts:

	Matthew	Mark	Luke	John	Acts	Romans	Corinthians
Genizah	1	-	-	1	1	-	-
Fayum	4	1	2	3	2	1	4
Oxy.	6	1	1	2	-	1	-
Hermop.	-	-	2	1	1	-	1
Panop.	-	-	2	4	1	-	1
Diospolis	1	-	2	1	-	-	-
Sinai	2	2	1	3	-	-	5

South of Hermoupolis, all three known findspots are monasteries. It looks once again as if there was not much read south of Hermoupolis, outside

[11] This includes the homily on Matthew *P.Schub.* 44.

monasteries, at this date. These figures are too small to tell us much about which towns were in contact with which others for purposes of religious reading, but it is worth noting that again much reading is taking place in the Fayum, while rather fewer texts than we might expect come from Oxyrhynchus or Hermoupolis.

Monasteries account for the majority of the books of which only one or two texts survive. Where their findspots are known, for instance, Galatians, Ephesians, Colossians, Thessalonians and Philemon, are found only at the White Monastery at Panopolis, while Philippians and Titus are found only at Panopolis and Sinai. Many of these texts have no provenance, so conclusions derived from these figures remain provisional; nevertheless, they are suggestively patterned.

It is difficult to assess the relationship between the Bible and literary culture in general. We do not know, for instance, how many texts of the Bible were used by only a few literate priests, perhaps for liturgical or catechetical purposes. It is possible that even quite large numbers of biblical finds involved rather few readers. On the other hand, far more people may have heard these texts than could read them, so some parts at least may have had far wider circulation than written remains can show. It may be that a better model for the reception of biblical texts in this period would be the reception of Attic drama in the fifth century, rather than the literary culture of the Roman world. My figures for and conclusions about biblical reading, therefore, should be taken with even more caution than the rest.

Is it possible, on the basis of all the figures collected so far, to compare the cultural profiles of the major findspots in the sixth century? Culture of all kinds is flourishing in the Fayum, despite waning links between the Fayum, Alexandria and other parts of Egypt, which we may perhaps see expressed in the slightly different reading or copying choices made by those who have left texts there. It is difficult to give exact figures where some texts are hard to identify, but there are over forty biblical texts from the Fayum in this period, over twenty other Christian texts and at least thirteen Classical texts, besides all kinds of other writings, from school exercises, legal, medical, magical and mathematical writings, to novels, panegyrics, astronomy and grammar.

At Oxyrhynchus there is not the same balance, but a notable overall bias in favour of classical texts, with around thirty identified classical fragments, compared with fewer than twenty biblical ones, around a dozen Christian writings and again, a large number of other writings. There clearly is a community reading Christian texts here, including some reading at a high level of sophistication, but Classical learning still outweighs Christian. The same pattern is discernible at Antinoopolis, where perhaps six biblical texts and eleven other Christian writings are outweighed by about twenty classical fragments. Again, there is a community reading Christian texts, and at a highly sophisticated level in some cases, but the emphasis is on classical learning. The same is true yet again at Hermoupolis, where fewer than fifteen biblical texts and only six other Christian texts are outweighed by around thirty classical fragments. Here too there is a significant number of other writings – around fifteen. These three towns are known as bastions of classical culture in Late Antiquity, and such bastions were apparently less interested in Christian reading than many less sophisticated areas.

The reverse is true at Tura, Aphrodite, the Herakleopolite, Panopolis and Sinai, which have only two or three classical texts among them, and large numbers of biblical texts and other writings. Apart from Tura, however, which has a collection of Origen, probably from one individual's library, these places have few sophisticated Christian writings – what they have is nearly all biblical or homilies. The Christian reading going on here is not, apparently, of a high intellectual order.

If the general picture emerging is one of classical learning in centres of learning, while Christian reading is more widespread but often happening on a lower intellectual level, the monastery of Mons Ephiphanius at Diospolis bucks the trend. Here, as in the Fayum, we have more of a balance: around fifteen biblical texts, seventeen other Christian writings, eight to ten classical fragments and around twelve other writings. Here, in a Christian context, classical learning was nevertheless valued.

Of the total of 1010 texts surviving from the 'long sixth century', some 131 are identified (some rather diffidently) as schooltexts. What can these texts in particular, tell us about the state of literate education, that *sine qua non* for literary culture in general?

Immediately striking is the range of findspots, where known. Among schooltexts overall, fewer than half of texts whose provenance is known, come from nome capitals, Greek cities or large towns. In the sixth century, the picture is very different: virtually all texts come from large urban centres or the still culturally active parts of the Fayum: thirteen from the Fayum, eleven from Hermoupolis, seven from Antinoopolis, three each from the Herakleopolite and Oxyrhynchus, two from Aphrodite and one from Dendera.

It seems that education has retreated from the smaller towns and villages to concentrate itself in the larger ones. Even elementary education seems to have withdrawn from small towns and villages. In general, schooltexts of elementary exercises are much more widely scattered through Egypt than those of grammar and rhetoric, which were always largely taught in the more cultured towns and cities. In the sixth century, the range of findspots is similar for all types of text.

Along with this, proportions of different types of schooltexts are rather different in the sixth century from what is observed in other periods. In the sixth century, for instance, rhetorical texts survive in a ratio to all texts of 1:10. For the Roman period the ratio is 1:21 and for the Byzantine period in general, 1:23. The implication of this seems to be that if fewer people, or people from a narrower geographical range, and perhaps social range, were receiving an education in the sixth century than at earlier times, a larger proportion of them was going on to the more advanced forms of study.[12] This would not be surprising if education had retained its popularity among the better off, but have become less common among poorer people. It is also possible that education had simply retreated into the larger cities, but in practice this would also mean that poorer people were not receiving it, since the poorer people were, the less they travelled.

There is some, slender evidence that education and literary culture go more closely hand in hand in the sixth century than in earlier times. In most centuries, schooltexts survive in fewer places than do texts, even literary texts, overall. In the sixth century, the fit is much closer: relatively few places have yielded evidence of literary culture without evidence of education too.

The content of education, on the other hand, has changed little. Reading and writing are still taught by means of letters, syllabaries, wordlists and passages of literature; numbers are still taught by number-lists, tables, and arithmetical and geometrical exercises. There are still grammatical tables using the same paradigm verbs, nouns and adjectives, and rhetorical exercises still begin with paraphrase and *ethopoiai*. Homer is still by far the most common author, followed by Menander and gnomologies and Euripides; as one would expect if education is relatively restricted to the better off, there is also a scattering of unusual authors – Demosthenes, Philemon and Palladius, along, of course, with the Old and New Testaments.[13]

The picture given here is of a state of education and culture somewhat shrunken in comparison with previous centuries. In many areas there is no sign of it at all. In a few it is clearly alive and flourishing. The range of what is read is still wide in some places, but not as wide as it has been in most.

The last exercise performed in my survey was to take a sample of 10% of texts, and consider the quality of the papyrus, the beauty, formality or informality of the handwriting, whether they have corrections or rewritings or other signs of scholarly use or long use over time, and all the other indi-

---

[12] Equivalent figures for the ratio of grammatical to all texts are 1:15, 1:13, 1:18. We should expect more people to study grammar than continue to rhetoric (Morgan, *Literate Education*, pp. 61-63).

[13] Morgan, *Literate Education*, *passim*; Cribiore, *Gymnastics*, *passim*.

cations by which papyrologists seek to discover how texts were actually used. Although the sample was small, the preliminary findings were striking. All the Christian texts which had been used in amulets, and most of the prayers, were in informal, often ugly, sometimes incompetent hands, suggesting that they had been copied by people who were not professional scribes and did not have a high level of education.[14] The biblical texts were overwhelmingly in rather large hands, fairly or clearly formal, occasionally calligraphic. Some of them had had lectional signs added and mistakes corrected, and in at least one, a second hand had written over faded letters, implying use over a long period.[15] This suggests texts which were copied for people (or communities) of varying wealth and education, for practical use, and the notably large hands of many would support my earlier suggestion that they were used liturgically. On the other hand, only one text in my sample looked likely to be that of a scholar: a fragment of biblical scholia in a quick, accurate hand with informal features.[16] There were none in the telltale informal hands with marginalia which scholars were still producing at this date. (One would, however, expect scholars' biblical texts in a large enough sample, since we know of so many Christian literates and teachers throughout the sixth-century world, so one should not conclude too much from this particular silence.)

Christian authors were mostly in good quality copies. One had cursive features, making it somewhat more likely to be a scholar's copy, though in my sample there were none with marginalia.[17] Pagan authors showed a much greater variety of hands, from high quality calligraphic texts to informal cursive texts, to poor quality copies, to schooltexts.[18] But there was a high proportion of informal hands, including four examples in my random survey of hands which had copied more than one pagan text: one, two Euripides texts; one, two Aristophanes texts; one, a Euripides and something unidentified; and one, a Callimachus with notes and an Apollonius Rhodius.[19] The last of these came from Oxyrhynchus and the first two from Antinoopolis.

These results tend to confirm the conclusions drawn from the data as a whole. Biblical material and some pagan literature is clearly being read by a

---

[14] E.g. *P.Oxy.* 1926, 1928, 2065.

[15] E.g. *P. Oxy.* 401, 846, 1169.

[16] *PSI* 1.65.

[17] *P.Oxy.* 2531; cf. *PSI* 13.1296.

[18] Calligraphic or good hands: *P.Oxy.* 1099, 1623, 3688, 4046, *PSI* 1.10, 13.1298, 1299; informal hands: *P.Oxy.* 1369, 3545, *P.A.t.* 1.24, 2.73, 74; poor hands: *P.Oxy.* 1013, 1614.

[19] Euripides: *P.Ant.* 1.23, 2.73; Aristophanes: *P.Ant.* 3.180, 211; Euripides *et al.*: *Archiv* 30.6, 18-29; Callimachus: *P.Oxy.* 2258.

variety of people and in a variety of contexts – with a probable emphasis on the liturgical in the case of the Bible. My sample showed up a high level of scholarly activity in pagan literature, both popular and more recondite, as well as calligraphic texts and schooltexts. There was, on the other hand, little sign of scholarly activity on the Bible or even on Christian authors, who seem mostly to be being transmitted in good quality copies which might be used for reading or for show. In general, my sample showed up rather few texts of a kind which in earlier times would be identified as 'show copies' – those in beautiful handwriting on top-quality papyrus. It is possible, of course, that people of no great wealth might keep a less beautiful papyrus for show. But most of these texts show signs of reading, such as wear of the materials, corrections, lectionals and so on. It seems more likely that this is a culture in which the copying and keeping of texts as cultural status-objects has diminished, and the texts which were owned were used, in one way or another. Most of them, however, do not show signs of having been read in a critical or scholarly way; what we have suggests a core of very sophisticated readers with a large penumbra of unsophisticated ones (or listeners).

The question remains, how the presence of archives affects our statistics. The answer must be that in the sixth century they affect them dramatically. A large proportion of all the literary texts we have seems to come from archives. Apart from the Dioscorus archive, which accounts for most of our literary texts from Aphrodite, archives account for most or all of our texts from Tura, Genizah, Sinai, Diospolis, and Panopolis (probably two archives). If we discount these archives, literary texts are even more thinly spread than they seemed.

How do we accommodate this into a model of literary culture in general? There seem to me at least three possibilities. The first is that culture was even more concentrated in a limited number of places than we thought – that the archives represent oases of culture in encroaching deserts of philistinism. The second is the opposite: that the accident of the survival of these archives – and Aphrodite was not even a city, but technically a village and in appearance a stylish but smallish town – implies that there were once many more archives and private libraries, and much more culture than is now visible or evidenced. The third is that greater numbers of educated people like Dioscorus grew up in Egypt than we have evidence for, but that most of them left, and the archives we have are evidence of the minority who stayed.

We have very little basis on which to choose between these. We do know of one or two highly educated individuals who left Egypt in the century before the days of Dioscorus and made enormously successful careers in Constantinople, such as the poet Cyrus from Panopolis, who became Prefect of Constantinople, and Orion the grammarian from Thebes who became an intimate of the Empress in Constantinople. Where there are one or two, the chances are that there were others who were not quite so distinguished and do not make it into the literary record.

Similarly, where we have found a handful of archives, it is hard to believe that there were not others which we have not found. But this, I think, does not add up to evidence of a much deeper culture than we thought. It is notable that in earlier centuries, despite the fact that we know of cultured individuals working in Egyptian towns, and also of public libraries, very few surviving texts can be grouped into archives. The fact that relatively many of our sixth-century texts can be so grouped still seems to me evidence of a lack of more widely dispersed culture, which makes the first suggestion more likely than the second: in the sixth century culture was if anything even more concentrated in particular places than we might think. Further, most archives containing literature come from the archives of places, such as monasteries, which we might expect to be centres of literacy, rather than from those of individuals, and most archives even of the sixth century, as from earlier Egypt, contain no literary texts.

Without Dioscorus, sixth-century Aphrodite would look rather poor in culture. With him, and when we add descriptions of the buildings and layout of his home town, it looks rather rich. But buildings and street plans, and material wealth, need not be indices of literary culture, and we should beware of assuming that they are. In the final analysis, one may still suspect that Dioscorus, in all his fascinating cultural complexity, was something of a rarity in his time.

THEOLOGY AND MEDICINE IN JUNDISHAPUR

CULTURAL CHANGE IN THE NESTORIAN SCHOOL TRADITION

Gerrit J. Reinink

Nestorian physicians of the city of Jundishapur in Khuzistan (Iran) played a fundamental role in the transmission of Greek medical learning to the Arabs in Baghdad (eighth-tenth centuries). As court-physicians of the 'Abbasid caliphs, as directors of the Baghdad hospital, in writing medical textbooks and commissioning translations of Greek medical works into Syriac and Arabic – in particular the works of Hippocrates and Galen – these Nestorian doctors contributed their share to the development of Islamic medicine in the Middle Ages.[1] According to general opinion these physicians are likely to have received their training in the famous medical academy and hospital of Jundishapur, institutions perhaps founded by the Persian shah Shapur I (241-272) or by Shapur II (309-379), and which reached their full flowering under the patronage of the shah Khusrau I (531-579).[2]

The existence of an old and noted medical school connected with a hospital in Jundishapur has, however, recently been called into question by some scholars. The latter would consign this success-story of Jundishapur to the realm of fancy, arguing that the tradition is no more than a 'myth' or 'literary invention' of the Middle Ages, one which was connected with the high reputation of certain Baghdad physician families which traced their origins back to Jundishapur.[3] In line with the opinion of these sceptics, David Lindberg has even concluded that 'there is no persuasive evidence for the existence of a medical school or a hospital at Jundishapur, although there seems to have been a theological school and perhaps an attached infirmary'.[4]

---

[1] Cf. Gutas, *Greek Thought*, pp. 118, 133.

[2] Cf. Gätje, *Grundriß*, p. 408. See also Elgood, *A Medical History*, p. 48; Hummel, 'Die Anfänge', p. 2; Schöffler, *Die Akademie*, pp. 33, 36, 40-41; Hau, 'Gondeschapur', p. 99; Frye, *The Golden Age*, pp. 22, 163.

[3] Nutton, 'From Galen to Alexander', pp. 12-13; Dols, 'The Origins', pp. 367-369; *idem*, 'Syriac into Arabic', pp. 49, 51; Conrad, Neve, Nutton, Porter and Wear, *The Western Medical Tradition*, p. 101; Jacquart and Micheau, *La médecine arabe*, p. 29.

[4] Lindberg, *The Beginnings*, p. 165.

Lindberg's conclusion may be carrying things a little too far. By 795 there certainly existed in Jundishapur besides a theological school a medical *bēt mardūtā*, or 'house of instruction'. The Syriac expression *bēt mardūtā* is synonymous with the Greek loan-word *'eskūlī* (σχολή) 'school' in Syriac, indicating a well-organised institution for the transmission of knowledge and the training of students.[5] The medical *bēt mardūtā* probably offered theoretical instruction of Hippocratic and Galenic medicine.[6] A letter of the Nestorian Catholicos Timothy I,[7] witness for the existence of this *bēt mardūtā* in Jundishapur, also contains interesting information about the socio-cultural position of this institution in the city. In this letter, which is addressed to Sergius, the Nestorian metropolitan of Jundishapur, Timothy informs Sergius of his having sent the student Gabriel to him for instruction (*mardūtā*), since the young man yearned for the study of medicine. Timothy, therefore, commissions the metropolitan to send Gabriel on to Zistaq and his *bēt mardūtā*.[8] It appears from this report not only that Zistaq was most likely the director of a special institution for medical training, but also that this school was supervised by the Nestorian metropolitan of Jundishapur. This last circumstance shows that the medical school was definitely a Christian institution, and it is, indeed, quite possible that the medical *bēt mardūtā* was connected with an infirmary belonging to the theological school in the city. Another interesting aspect of Timothy's letter is the fact that the medical school in Jundishapur obviously enjoyed a certain reputation. For why should Timothy send one of his students to the medical school in Jundishapur if there were comparable training facilities in the old partriarchal city of Seleucia-Ctesiphon near Baghdad?[9] The same Timothy

---

[5] Cf. Payne Smith, *Thesaurus Syriacus*, I, col. 490. For the use of the word *mardūtā* for scholastic instruction, cf. for example the teaching activity called *mardūtā* in the school (*'eskūlā*) of the pagan philosophers in Alexandria by Barhadbeshabba, teacher in the school of Nisibis about 582-595 and the author of the *Cause of the Foundation of the Schools*: Scher, *Mar Barhadbšabba*, pp. 375-376.

[6] The Syriac translations of Galen's works by Sergius of Resh'ayna (died 536) were already in the first half of the sixth century in circulation among Syrian scholars of the Persian empire; cf. Hugonnard-Roche, 'Note', pp. 123-125. The Nestorian Catholicos Aba I (540-552) may also have played a prominent part in introducing Greek medical learning into the Persian church; cf. Brock, 'From Antagonism to Assimilation', p. 22.

[7] For the life and the works of the Catholicos Timothy I (727/8-823), see Putman, *L'église et l'islam*, pp. 13-23. For Timothy's role in translations of Greek philosophical works into Arabic, see Brock, 'Two letters', pp. 233-246.

[8] Braun, *Timothei patriarchae I epistulae*, pp. 71/108.

[9] In 780 Timothy transfered the patriarchate from Seleucia-Ctesiphon to Baghdad (founded in 762); cf. Tisserant, 'L'église nestorienne', col. 194; Putman, *L'église et l'islam*, p. 128; Allard, 'Les chrétiens à Bagdād', pp. 377-378.

provided the amount of 20.000 *zūzē* for the building of a *bimaristan* 'hospital' (=Pers. 'place for the sick') in Seleucia-Ctesiphon, which may indicate that at that time such an institution did not yet exist in the old patriarchal city.[10]

We know that there was a Nestorian school of theology in Jundishapur, at least by the end of the sixth century, in which, after the elementary instruction of reading and writing in the Syriac language, the theological tenets of the Church, biblical exegesis and Aristotelian logic could be studied.[11] Moreover, the theological school stood under the superintendence of the metropolitan, and the metropolitan Sergius, the addressee of several of Timothy's letters,[12] may have functioned as one of its teachers and even as director.[13] The fact that there were in Jundishapur two Christian schools with separate training-courses, one for the schooling of the clergy, the other for the training of medical practitioners, both institutions being in the charge of the metropolitan, makes the assumption plausible that the situation in Jundishapur may not have differed much from that in Nisibis, the Nestorian metropolis of North Mesopotamia.

In the middle of the sixth century, under the directorate of Abraham of Bet Rabban (from about 510 until about 569), the famous theological school of Nisibis, the oldest East Syrian teaching institution on a larger scale,[14] had received a hospice for the nursing of students who had contracted an illness. The foundation of this *xenodocheion*, as this institution was called,[15] formed part of an extensive building programme necessitated by the strong growth of the number of students.[16] It was no longer possible to tend to the sick students in the cells,[17] and it was considered undesirable that they enter the city and be exposed to the greed of the citizens.[18] It is possible that the

[10] See the unpublished letter 44 to Sergius, summary in Bidawid, *Les lettres*, p. 35; cf. Dols, 'The Origins', pp. 379-380.

[11] Cf. Schweigert, 'Die Theologenschule', pp. 185-187; for early Christianity in Khuzistan, see *idem*, *Das Christentum in Hūzistān.*

[12] Cf. Bidawid, *Les lettres*, pp. 68-72.

[13] As is suggested by Schweigert, 'Die Theologenschule', p. 186.

[14] Vööbus, *History* (for Abraham's directorate, see pp. 134-210).

[15] For the origin of the Byzantine *xenodocheion*, 'place for strangers', which came to mean 'house for the poor' in Christian tradition, see Miller, *The Birth*, pp. 8-9, 26-27. The Greek loanword was used in Syriac in Mesopotamia and Syria to designate in particular an institution for the care of the sick; cf. Dols, 'The Origins', pp. 372-373.

[16] Cf. Vööbus, *History*, pp. 143-147.

[17] Canon XI of the Canons of the director Narsai (set up in 496) prescibed that the sick should be nursed in the cells by their brothers; cf. Vööbus, *The Statutes*, p. 80.

[18] Cf. Nau, *Barhadbešabba*, pp. 622-623 [134-135]. This Barhadbeshabba, author of the *Ecclesiastical History* (written after 569), probably is to be distinguished from

foundation of this *xenodocheion* was supported by Khusrau I. According to the anonymous author of Books VII-XII of Pseudo-Zacharias' *Ecclesiastical History* (probably a monk from North Mesopotamia, who wrote around 569)[19] Khusrau I founded the first *xenodocheion* in the Sasanian empire on the advice of the Christian physicians at his court.[20] Among those physicians, who were highly esteemed by the shah, were Tribunus, a Byzantine *archiatros*, and the Nestorian Catholicos, Joseph.[21] Pseudo-Zacharias also mentions the names of Birway, Qashway and Gabriel, a Christian from Nisibis. Qashway is also mentioned by Barhadbeshabba in his report of the life of Abraham of Bet Rabban. He was the Christian 'physician of the king' Khusrau, who provided Abraham with a site in Nisibis for the new buildings of the school.[22] It is likely that Qashway was also involved in the physicians' appeal to the shah for the building of a *xenodocheion* and that this hospice was none other than the one built by Abraham in Nisibis.

If this conclusion is right, we have some more information about the earliest provisions of the *xenodocheion* in Nisibis. Pseudo-Zacharias reports that Khusrau supported the *xenodocheion* by providing many supplies and twelve physicians. The presence of such a number of professional doctors created favourable conditions for the training of medical students. We do not know when the *xenodocheion* in Nisibis started training such students, but it is certain that training facilities existed there by 590, when Henana was director of the theological school.[23] Three canons of the school of Nisibis set up during the teaching of Henana concern the *xenodocheion* and medicine. Canon I regulates the duties, responsibilities of the *xenodochos*, the curator of the hospital, and the sanctions which are to be laid upon him, if he break the rules.[24] Canon XIX prohibits the students of theology to live together with the students of medicine 'in order that the books of the craft of the world should not be read with the books of the holiness in one

the author of the *Cause of the Foundation of the Schools* (above, note 5); cf. Reinink, 'Edessa grew dim', p. 81, and note 15.

19 Cf. Brock, 'Syriac Historical Writing', pp. 4-5.

20 Brooks, *Historia ecclesiastica*, pp. 217-218 /147; English transl. by Hamilton and Brooks, *The Syriac Chronicle*, pp. 331-332.

21 For Tribunus, see Evans, *The Age of Justinian*, p. 166. For Joseph (552-567), who had studied medicine in the Byzantine empire and who was elected as Catholicos by the intervention of Khusrau I, see Labourt, *Le christianisme*, pp. 192-197; Moffett, *A History*, pp. 232-233; Vööbus, *History*, pp. 159-160.

22 Nau, *Barhadbešabba*, p. 623 [135].

23 For Henana's directorate, see Vööbus, *History*, pp. 234-317; Reinink, 'Edessa grew dim'.

24 Vööbus, *The Statutes*, pp. 92-93.

light'.[25] Canon XX prohibits students, who have left the study of theology for the study of medicine without having a good testimony, to attend the lectures in the theological school.[26] According to the *Chronicle of Seert*, Babai the Great studied the books of medicine in the hospital, and the ecclesiastical sciences in the school of Nisibis for fifteen years.[27] If we may trust this report,[28] Babai would thus have studied medicine in the *xenodocheion* somewhere in the sixties of the sixth century.[29]

Consequently, we may conclude that by the end of the sixth century there existed in Nisibis a *xenodocheion* attached to and organisationally closely connected with a theological school, in which not only was nursing available for sick students but theoretical medicine was also cultivated. But what was the position of the metropolitan of Nisibis vis-à-vis both institutions? The proems and ratifications of the canons of the school of Nisibis clearly show that the setting up of canons was authorised by the metropolitan and that they required the metropolitan's sanction through his seal and signature.[30] Since the canons of the school included regulations concerning the administation of the *xenodocheion* and also the relation between theological and medical students and studies, both institutions were equally and ultimately under the authority of the metropolitan of the city.[31]

That the situation in Jundishapur probably reflected that in Nisibis may be indicated by another circumstance. The canons of Henana suggest that in Nisibis students of theology were in close contact with students of medicine, that students, moreover, easily switched from one study to another, and that it was not unusual for students of medicine to attend the lectures in the theological school. Thus a number of Nestorian theologians and clerics were also medically trained, and, as we have noted above, there are examples of high-placed ecclesiastical authorities who practised medicine. Conversely, there were many physicians whose daily pursuits were determined by the practice of the medical profession, and who possessed in addition a

[25] *Ibidem*, pp. 100-101.

[26] *Ibidem*, p. 101.

[27] Scher, *Histoire nestorienne*, pp. 530-531 [210-211].

[28] According to Isho'denah of Basra's *Book of Chastity* (c.850) Babai studied doctrine and (biblical) commentaries for fifteen years and practised as a *malpānā* 'teacher' in the *xenodocheion* in Nisibis; Chabot, *Le livre*, pp. 22/25. Isho'denah does not state explicitly that Babai studied medicine or practised as a teacher of medicine in the *xenodocheion*.

[29] Perhaps in or shortly after 571 Babai entered the monastery of mount Izla to the north of Nisibis, newly founded by Abraham of Kashkar; cf. Chediath, *The Christology of Mar Babai*, pp. 4-5.

[30] Vööbus, *The Statutes*, pp. 51-52, 55, 58-60, 91-92, 105.

[31] Cf. also Vööbus, *History*, pp. 269-275.

considerable knowledge of theological and philosophical matters. We know that some Nestorian physicians from Jundishapur had received an education in the theological school of the city. Gabriel bar Boktisho' is one example – a member of the famous Boktisho' family of physicians from Jundishapur,[32] who served from 805 as the personal physician to the caliph Harun ar-Rashid,[33] and who before 794/5 studied biblical exegesis, the works of the churchfathers and philosophy in the theological school of Jundishapur.[34] We have conclusive evidence that a member of another well-known family of physicians in Jundishapur, who lived in the ninth century, was trained as theologian and as physician, and may have been active as *malpānā* 'teacher' in both the theological and the medical school. His Persian name is Seharbokt, his Christian name Seliba Zeka 'the Cross has overcome'.

Seharbokt's father was the doctor Mesargis, who is probably identical with the Christian doctor Masarjawaih from Jundishapur known in the Arab tradition as a contemporary of the poet Abu Nuwas (died about 815).[35] In the Arab tradition Seharbokt is known, under his Arabised name Saharbuht b. Masarjis, as the author of a commentary on the '*Kunnash*', a medical handbook written by Jurjis.[36] Arab sources mention other members of the Mesargis family: 'Isa b. Masarjis, who was probably a brother of Seharbokt, and 'Isa b. Saharbuht, the son of Seharbokt, known as the translator of Hippocratic works from Syriac into Arabic.[37] Since the discovery of many quotations from the writings of 'the priest and teacher' Seharbokt bar Mesargis in the *Gannat Bussame*, a Nestorian exegetical compilation from the tenth century,[38] we know that Seharbokt also wrote extensive biblical commentaries on the Gospels of Matthew, Mark, Luke and John, the Acts of the Apostles, and the Epistle to the Romans.[39] Since the author of the *Gannat Bussame* used Seharbokt as a main source for the exegesis of the above-mentioned New Testament writings, we can form a rather clear notion of the literary character of Seharbokt's biblical commentaries.[40]

---

[32] Cf. Ullmann, *Die Medizin im Islam*, p. 111.
[33] *Ibidem*, p. 109.
[34] Schwaigert, 'Die Theologenschule', p. 186.
[35] Sezgin, *Geschichte*, p. 224.
[36] *Ibidem*, p. 242. Jurjis is identified with George bar Gabriel bar Boktisho' (second half of the eighth century) by Dietrich, *Medicinalia Arabica*, p. 67, note 2; Sezgin, *Geschichte*, p. 210, rejects this identification.
[37] Sezgin, *Geschichte*, pp. 242-243.
[38] For the date of the *Gannat Bussame*, see Reinink, *Gannat Bussame*, pp. IX-XI (transl.).
[39] Reinink, *Studien*, pp. 113-115.
[40] *Ibidem*, pp. 277-278.

Quite in line with his contemporaries, the Nestorian exegetes Theodore bar Koni,[41] Isho' bar Nun[42] and Isho'dad of Merv,[43] Seharbokt's biblical commentaries are typical representatives of the encyclopedic works based on the long and complicated school tradition of East Syrian biblical exegesis.[44] In this respect Seharbokt's originality, as compared with the abovementioned compilators, is restricted to the choice of his sources, the Syriac and Greek exegetical authorities of the past, the selection of the materials that he took from these sources, and the way in which he organised these materials in his own composition.[45] In one respect, however, Seharbokt stands out in a most remarkable way from his colleagues. His biblical exegesis is imbued with medical knowledge. Wherever a scriptural passage leaves scope for medical discussion, Seharbokt seizes the opportunity to dwell on medical themes and to display his medical knowledge. However, Seharbokt's interest in medicine in these biblical commentaries is more than a display of encyclopedic erudition alone. Theoretical medicine has become in Seharbokt's biblical exegesis a veritable ancillary science for the explanation of Scripture, and this circumstance marks a drastic change in the Nestorian theological school tradition. Whereas canon XIX of Henana's statutes of the school of Nisibis draws a sharp distinction between religious and medical studies, and the Nestorian theological tradition before the ninth century in general shows no special interest in theoretical medicine, in Seharbokt's theological works Galenic medicine appears as an instrument applicable in the elucidation of biblical texts.[46] The explanation for that must be sought in the specific socio-religious situation in Jundishapur on the one hand and in the general cultural climate in the 'Abbasid caliphate during the ninth century on the other.

Since he spent his childhood in Jundishapur, and since he in his later life prosecuted the profession of teacher of biblical studies in that city, Seharbokt must have been closely connected with the theological school of

---

[41] For Theodore's *Scholion* (completed around 791/2), see Van Rompay, 'Development', p. 566; McCullough, 'Early Syriac Commentaries', pp. 22-23.

[42] For Isho' bar Nun's *Selected Questions* (written before 828), see Van Rompay, 'Development', p. 567; McCullough, 'Early Syriac Commentaries', p. 98.

[43] For Isho'dad of Merv's *Commentaries on the Old and New Testament* (ca.850), see Van Rompay, 'Development', pp. 569-570; McCullough, 'Early Syriac Commentaries', pp. 23-24.

[44] Cf. Van Rompay, 'La littérature exégétique syriaque', pp. 221-235. For the literary character of Seharbokt's biblical commentaries, see Reinink, *Studien*, pp. 123-126; *idem*, 'Seharbokt', pp. 77-80.

[45] For a comparison between Seharbokt's and Isho'dad's exegesis, see Reinink, *Studien*, pp. 127-132.

[46] Reinink, 'De invloed', pp. 31-42.

Jundishapur, and he probably had received his theological education at that same institution. As regards his medical knowledge and training, Seharbokt will undoubtedly have learned much from his father, the physician Mesargis, who very probably maintained close relations with the medical *bēt mardūtā* in that city. The fact, therefore, that his son Seharbokt is known not only in the Syriac tradition as the author of exegetical handbooks, but also in the Arabic tradition as the author of a medical handbook, strongly suggests that Seharbokt officiated as a teacher in both institutions.

However, it is not only the fact that Seharbokt was a cleric who was trained in theology and medicine that explains his unique position as biblical scholar as compared with the exegetes of his time. An additional explication must be sought in the connections that existed since 765 and throughout the ninth century between medical circles in Jundishapur and the Arab-Islamic milieu and the 'Abbasid court in Baghdad.[47] Ninth-century Baghdad was the scene and the focal point of what be may called a cultural revolution in the Arab-Islamic tradition: the Graeco-Arabic translation movement which took place from the eighth to the tenth centuries.[48] In this movement Syriac-speaking Christians played a fundamental role as mediators of (Late) Antique Greek culture to the Arabs, through the provision of Syriac-Arabic and Graeco-Arabic translations.[49] This activity of the Syriac-speaking Christians had important effects on the Syriac tradition as well, since it gave a new impulse there to the revival of Greek sciences and generated a new influx and reception of pagan Greek scientific works in the Syriac Christian culture.[50]

As has recently been emphasised by Dimitri Gutas, during the reign of al-Ma'mun (813-833) the translation movement became an instrument of the caliph's policies of politico-religious propaganda, through which the caliph presented himself as the champion of Islam. Al-Ma'mun's aggressive policy against the Byzantines was attended by an ideology according to which it was not Christianity but Islam that was the natural and true heir of

---

[47] In this year George bar Gabriel bar Boktisho', director of the hospital in Jundishapur, was called to Baghdad by the caliph al-Mansur; cf. Gutas, *Greek Thought*, p. 118.

[48] O'Leary, *How Greek Science Passed to the Arabs*, pp. 146-175; and in particular, see now Gutas, *Greek Thought, passim.*

[49] O'Leary, *How Greek Science Passed to the Arabs*, pp. 155-171; Gätje, *Grundriß*, pp. 420-431; Gutas, *Greek Thought*, pp. 136-141; Conrad, Neve, Nutton, Porter and Wear, *The Western Medical Tradition*, pp. 104-110. For the historical and cultural background of the Syriac translation activity, see Brock, 'From Antagonism to Assimilation'.

[50] Baumstark, *Geschichte*, p. 227; Baum and Winkler, *Die apostolische Kirche*, pp. 61-63.

ancient Greek science and wisdom. Thus, 'a cultural attitude' was generated, 'in the centuries after al-Ma'mūn, which could easily regard as factual the tendentious "histories" of Greek philosophy and science in which the Christians are portrayed as having prohibited philosophy and medicine (and the sciences) because of their fear of reason, while the Muslims are pictured as champions of the truth'.[51] One of the main reasons why Seharbokt made medicine the ally of theology in his biblical commentaries and teaching may have been his intention to demonstrate that his Christian religion was in perfect harmony with Greek science, and that Scripture in no way conflicted with reason and Greek wisdom.

Apologetical tendencies appear clearly in a number of Seharbokt-traditions in the *Gannat Bussame*. Of course, Christian demonology and miracles, as these occur in the biblical writings, are by no means denied by Seharbokt or *replaced* by 'rational' medical explanations. Religion and science are presented as being complementary to each other. Thus, Christ is not simply an exorcist, who heals people by casting out demons, but He is a 'true doctor' by virtue of His healing of real physical disorders caused by demoniac activity or other circumstances.[52] In other instances, traditional demonological explanations of diseases are amplified by medical explications. Thus, Seharbokt, explaining the Syriac name of the epileptic boy's demon in Matth. 17:15: 'Lord, have mercy on me; my son has a "son of the roof" (*bar 'eggārā*), and he is gravely ill', places a medical physiological explanation of the 'son of the roof' side by side with a purely demonological explication.[53] Seharbokt knows the traditional Nestorian exegesis, which explains 'son of the roof' as the name of the demon who is so called because his activity is connected with the roof (he throws his victims from the roof or from other steep places),[54] and he puts this opinion into the mouth of the boy's father.[55] Seharbokt, however, is the only Nestorian exegete who also advances a medical explication, according to which 'son of the roof' is

---

[51] Gutas, *Greek Thought*, pp. 83-95, esp. 95.

[52] See Reinink, 'De invloed', p. 36.

[53] In the Syriac Bible the Greek verb *selēniázetai*, '[for he] is lunatic', *i.e.* an epileptic (Matth. 17:15), is translated by '[he has] a *bar 'eggārā*', 'son of the roof'. *Bar 'eggārā* is the name of the demon of epilepsy known in Sumerian as Lugul-urra, 'Lord of the Roof'; cf. Stol, *Epilepsy in Babylonia*, pp. 16-19.

[54] Reinink, 'Der Dämon', p. 110.

[55] *Gannat Bussame*, commentary on Matth. 17:14-15, fol. 730^v. I quote the still unpublished parts of the *Gannat Bussame* according to the fols. of the manuscript Louvain, CSCO, syr.14. For the description of this MS, see Reinink, *Gannat Bussame*, pp. IX-X (ed.); for the textual history of the *Gannat Bussame*, see *ibidem*, pp. XVII-XXI.

the name of the boy's illness.[56] The illness is called 'son of the roof', Seharbokt says, because the afflicted one (*metnassyānā*) hallucinating thinks that a spirit from the roof rushes at him, in order to harass him.[57] The illness is still due to demoniacal activity, since the word *metnassyānā* has the specific meaning of being afflicted by a demon,[58] but the demon causes a real disease of the brain, which affects the visual faculty and makes the patient suffer hallucinations.[59] Highly apologetical also is Seharbokt's commentary on Mark 16:17-18, where Christ prophesies that His followers shall cast out demons in His name, speak with new tongues, take up serpents, drink deadly poison without being hurt, and lay hands on the sick, so that they shall be healed. These wonders have been performed by Christians throughout the ages and generations, Seharbokt says, whereupon he gives a number of examples from the past and the present.[60] One of the stories related by Seharbokt has a clearly polemic note addressed against the Muslims. Some time ago in Jundishapur there was a monk, Seharbokt relates, who behaved as if mad. When he was invited by the Muslims to come with them to the mosque in order to pray with them (so that he might be healed), he promised them to come the following morning. This delay gave the monk the opportunity to go to the desert first and to catch a number of serpents, which he hid in his tunic. The next morning, in the mosque, the monk released the snakes, and all Muslims fled from the mosque, whereupon the monk spoke: 'I do not pray with somebody who flees before this, since he does not fear for God alone'. The wonders pro-phesied by Christ are therefore real, and show that the Christian believers at all times possessed the true fear of God and were protected by their faith. The prodigious healings by Christians continue, moreover, throughout the ages and today in 'this difficult time' they reveal the truth of Christianity: such are the healings by prayer, the imposition of hands, the signing with the sign of the cross and the anointing with the *hnānā* (a compound of oil, dust, and water mixed with the relics of the saints or with earth of holy places).[61] But Christians are also successful, Seharbokt continues, in healing diseases by the prescription of various diets

[56] For the medical explications in the commentary of the Syrian Orthodox writer Dionysios bar Salibi (died 1171), see Reinink, 'Der Dämon', p. 109, note 17.
[57] *Gannat Bussame*, commentary on Matth. 17:14-15, fol. 730^v.
[58] The Ethpa. *'etnassī* has the specific meaning of being possessed by a demon; cf. Payne Smith, *Thesaurus*, II, col. 2389.
[59] Cf. Budge, *Syrian Anatomy*, pp. 74/79. For the Galenic background, cf. Meyerhof, 'Die Augenheilkunde', pp. 257-268; cf. also Degen, 'Ein Corpus', pp. 116-117, n. 16, and Gignoux, 'Le traité', pp. 725-733.
[60] *Gannat Bussame*, commentary on Mark 16:17-18, fols. 643^r-644^v.
[61] Cf. Payne Smith, *Thesaurus*, I, cols. 1315-1316.

and foods,[62] just as, according to antique Galenic medicine, the balance of the humours in the human body could be restored by eating the right foods.[63] Even Christ Himself initiated in the Last Supper a healthy diet: the dish with boiled food (*būšālā*) into which He dipped the bread (John 13: 26) did not contain the bitter plants used by the Jews,[64] but rather beetroot, raisins and nourishing vegetables.[65] Following Galen's *On the Powers of Foods*, Seharbokt stresses the purgative properties of beetroot and raisins.[66]

No less than Islam, the Christian religion is compatible with and open to medical science. As a doctor in the theological school in Jundishapur, Seharbokt teaches his students that Greek science, Greek medicine in particular, is a good additional expedient for the interpretation of Scripture, just as the practice of religion and that of medicine can go hand in hand. In his commentary on Mark 5:26, where it is said concerning the woman who for twelve years suffered from menstrual fluxes that 'she suffered much at the hands of many physicians, and had spent all that she had and was not helped at all, but rather became worse', Seharbokt feels called upon to make an apology for the medical profession.[67] First of all, Seharbokt says, we should not take the words 'she suffered' (*seblat*) in the sense of suffering from some physical or mental disorder caused by the physicians. For 'she suffered' in reality from a disease in the belly and the intestines, and she suffered the humiliations from the people because of this illness. The words 'she suffered' (*seblat*) have rather to be explained as 'she endured' (*saybrat*), for she endured many things from pretended and incompetent physicians puffed up with boasting and big talk. 'I feel confident', Seharbokt continues, 'that a skilled physician is beneficial – and I call a *skilful* (physi-

---

[62] As examples Seharbokt mentions Boktisho' d-Miles who used to cure fevers by prescribing the eating of walnuts, the recluse Adarshabur who used to purge and heal people suffering from dropsy by prescribing the eating of mouldy bread, and his own recovery from tertian fever (Seharbokt uses the Greek loanword *tritaios*) by eating salt fish and honeycomb. Boktisho' d-Miles may be a (otherwise unknown) member of the famous Nestorian Boktisho' family of phycisians from Jundishapur. Seharbokt connects Adarshabur with the monastery of Rabban Shabur in Khuzistan (cf. Fiey, *Assyrie chrétienne*, p. 208, n.4).

[63] Grant, *Galen on Food*, pp. 8-10.

[64] Seharbokt relates these bitter plants to Ex. 12:8 and the Jewish custom of eating boiled bitter herbs at Passover in memory of the acrimonies which the Hebrews suffered in the desert for forty years. He identifies these bitter herbs with wild cucumber, fennel and bitter lettuce.

[65] *Gannat Bussame*, commentary on Matth. 26:23, fols. 481^{v}-484^{v}.

[66] Kühn, *Claudii Galeni Opera Omnia*, p. 582 (raisins), p. 630 (beet); transl. Grant, *Galen on Food*, pp. 120, 140. For the Syriac translations of this work of Galen, see Degen, 'Galen im Syrischen', pp. 139-140.

[67] *Gannat Bussame*, commentary on Mark 5:26, fol. 744^{v}.

cian) the God-fearing (physician) who is trained and skilful in medical matters and likewise at all times resorts to God, asking Him to make his (i.e. the physician's) treatment succesful'. The healing art and religion are not conflicting, but rather complementary entities, since a wise and skilful doctor is a pious man who knows that his medical treatment is favoured through praying for divine assistance. It is most likely that Seharbokt is speaking here from his own experience, and that he was not only a teacher of theology and medicine, but also a medical practitioner like his father.

The quotations from Seharbokt's New Testament commentaries in the *Gannat Bussame* belong to the rare remaining witnesses of the important cultural change which took place in ninth century Nestorian school tradition. Owing to the fact that the *Gannat Bussame* became the Nestorian standard commentary on the biblical lectures of the ecclesiastical year,[68] the symbiosis between theological and medical sciences became a regular constituent in the Nestorian school tradition right up to modern times.

[68] Cf. Reinink, *Gannat Bussame*, pp. XI-XXI (transl.).

CAROLINGIAN BISHOPS AND CHRISTIAN ANTIQUITY

DISTANCE FROM THE PAST, CANON-FORMATION, AND IMPERIAL POWER

Michael E. Hönicke Moore

In what follows, I wish to show that the establishment of a canon of patristic authors, in the Carolingian period, reflected a new vision of the Christian past, serving as the historical backdrop for an empire making sweeping religious claims. In favour of an emphasis on authors of the Christian past, the independence of bishops was compromised, and their reliance on local traditions was held up against a universal, ancient standard. This was reflected in the use of the Fathers as a source of canon law beginning in the late eighth century.

Patristic sources were rarely cited in Gallican and Merovingian councils (314-751), despite their constant claim of faithfulness to the 'ancient fathers'. As will be seen, the Council of Frankfurt, held in 794, was a turning-point. Thereafter, Carolingian councils quoted extensively from the early Christian authors, and found in them the resources of a new legislative programme. By studying this problem, it may be possible to turn from a passive concept of tradition toward a theme of appropriation and transposition of the past.

It may appear evident to us who 'the Fathers' were: those authors of Christian antiquity, such as Jerome and Augustine, whose works dominated the early Middle Ages. As Heinrich Fichtenau put it, 'all creative effort was replaced by the wish to pass on a tradition and to hold fast to the authority of the earlier Christian authors'.[1] Fichtenau thus described tradition as the continuous passing-on of a stable cultural inheritance. However, discontinuity and profound change in this tradition call for further reflection.

The flexibility and ambiguity of the term 'the Fathers' highlights the problem. In the First Epistle of Clement, the Apostles were called Fathers,

---

My thanks to Justine Firnhaber-Baker, David Nirenberg, Osvaldo Pardo and Michaela Hönicke Moore for their help and comments on this paper.

[1] Fichtenau, *The Carolingian Empire*, p. 98.

as a special mark of veneration.[2] The term was gradually extended to all bishops – while during the doctrinal disputes of the fourth century, the term 'Father' came to refer to those authors who defended true doctrine.

There was no list of such authors, and the term was not carefully restricted. Moreover, modern use of the term does not correspond to ancient or medieval usage. The Fathers, scholars often say, were a group of authors whose works had special authority – prized for their orthodoxy, their saintly lives, the broad acceptance of their writings, and their antiquity.[3]

Indeed, throughout the period addressed here, roughly 400-800 AD, the word 'ancient' carried a strongly positive emotional connotation. And yet the creation of the Christian tradition involved the evaluation and marshalling of writings from the past, and not merely a devoted passing-on. I would suggest further that the formulation of a tradition arises within a mental picture of the past, which affects how that tradition is assembled and used.

The writings of the Christian past did not arrive in early medieval Europe with implacable momentum or universal recognition. Translation, re-edition and even reattribution of patristic works were constant features of their transmission.[4] Many of the books which the Carolingians later recuperated in their efforts at library-building had barely made it across the sixth and seventh centuries. This was true even of the *Confessions* of St. Augustine.[5]

There were a handful of bibliographic guides to the authors of Christian antiquity. Such works did not seek to establish every acceptable authority, or to exclude all others. This is especially true of the influential *De libris recipiendis et non recipiendis*, a list of accepted authorities and condemned books compiled in about 520.[6] Attributed to Pope Gelasius I (492-496), the

---

[2] 1 Clem. 62.2.

[3] Pelikan, *The Christian Tradition*, p. 336; Altaner, *Patrologie* p. 3ff. These criteria (orthodoxy, antiquity, sanctity of life, acceptance by the Church) are discussed at greater length in Congar, *La Tradition*, vol. 2, pp. 192-207.

[4] This has long been recognised in scholarship on the Christian tradition. See for example: Baur, 'L'entrée littéraire', and Bardy, 'Éditions et rééditions'; Quasten (*Patrology*, vol. 4, pp. 195-212).

[5] Wilmart, 'Les manuscrits des "Confessions"'.

[6] Cf. Von Dobschütz, *Das Decretum Gelasianum*. In an overview of the literature, Von Dobschütz dated the work to the first half of the sixth century (pp. 334-348). Wurm argued that, in a first edition of the canon-collection known as the *Dionysiana*, Dionysius Exiguus included 50 canons from the Apostolic Canons. In a later edition he excluded them on the authority of the so-called 'Gelasian Decretal' (Wurm, *Studien und Texte*, pp. 24-26). This second edition was made sometime around 520 (Gaudemet, *Sources du droit*, pp. 134-137). The preface of the second redaction indicated the change (Maassen, *Geschichte*, pp. 964-965). The first edition

book issued a warning about authors such as Rufinus and Origen, whose works could only be accepted in part, and a list was provided of forbidden heretical authors.[7]

Naming only twelve authors, from Cyprian to Prosper of Aquitaine, whose books were recommended, the compiler gestured vaguely to include 'the works and tractates of all the orthodox fathers who deviate in nothing from fellowship with the church of Rome'.[8] The *Decretal* only pointed to a north star of orthodoxy.

A second major guide was Jerome, whose work *On Illustrious Men* was widely diffused later in the Middle Ages. Although it was not Jerome's intention to provide a list of authoritative writings, it later came to serve in this capacity for Carolingian librarians.[9] Jerome's work was expanded by Gennadius of Marseille (d. 492-505), and in this form was very influential.[10] Mathisen has noted that Gennadius was part of a movement to assemble legal, liturgical and patristic authorities, with the dissolution of Roman power in Europe.[11] Isidore of Seville (c. 560-636) also wrote in this genre, and like Jerome used his *De viris illustribus* to assign praise and blame to authors of the past.[12]

---

of Dionysius may be found in Strewe, *Die Canonessammlung*. Nevertheless, McKitterick has argued that the pseudo-Gelasian *Decretal* should instead be assigned to c. 700 on the strength of its manuscript tradition, and an appropriate context in a north Frankish or Burgundian centre of the period (McKitterick, *The Carolingians*, pp. 202-203). Nothing internal to the work offends an early sixth-century date.

[7] Dobschütz, *Decretum Gelasianum*, pp. 44-45 and 58-60.

[8] *opuscula atque tractatus omnium patrum orthodoxum, qui in nullo a sanctae Romanae ecclesiae consortio deviarunt ...* (Dobschütz, *Decretum Gelasianum*, pp. 36-38).

[9] Vessey ('The Forging', pp. 508-509) points out that Augustine wished that Jerome had made his work a guide to orthodox *loci*. Jerome claimed to take Suetonius as a model for his work: *Tranquillum sequens* (*Praef.* 1). Cf. Ceresa-Gastaldo, *Gerolamo*, p. 56. But this may be a learned gesture rather than a real indication of his inspiration, which was probably Eusebius: Dekkers, 'Pour une histoire', pp. 54-57 and Feder, *Studien*. On Suetonius's *De viris illustribus*, of which only fragments survive, see Cizek, *Histoire*, pp. 257-258.

[10] Gennadius provided extensive lists of books written by the authors he named *De viris inlustribus* (Richardson, *Texte und Untersuchungen*). See also *Gennadius als Litterarhistoriker* (Czapla, *Kirchengeschichtliche Studien*); Pricoco, *Storia letteraria*; Morin, 'Le *Liber dogmaticum*'; Mathisen, *Ecclesiastical Factionalism*, p. 242; Altaner, *Patrologie*, p. 425; and De Ghellinck, *Patristique et moyen âge*, vol. 2, pp. 246-248.

[11] Mathisen, 'The "Second Council"', pp. 511-513 and 544-545; similarly Le Bras, 'Notes'.

[12] Codoñer Merino, *El "De viris illustribus" de Isidoro de Sevilla*. On the

Ildefonse of Toledo (657-667) produced yet another *On Illustrious Men* as an act of piety toward his predecessors.[13] Half of his illustrious men were bishops of Toledo, alongside Isidore of Seville and other Spanish bishops. At the beginning stood Asturius of Toledo (fl. c. 400), and at the end, Ildefonse's predecessor, and uncle, Eugenius (646-657).[14] Literary accomplishment was not the standard: Asturius, Bishop of Toledo 'established the works of virtue more by the example of his life than with the pen of authorship'.[15] Bishop Aurasius 'taught with words more than with writing'.[16] Ildefonse wanted to show the coherence of a moral tradition, expressed in written works, but just as authentically in the spoken word and actions.[17] His illustrious men formed a bridge to Christian antiquity that was firm and easy to cross, rooting the local past of Toledo.

Despite the scarcity of evidence, it is possible that these works were used early on as guides to gathering books. At any rate, these guides did not come into their own until the Carolingian period.[18]

The writings of the Christian past were gathered in libraries whose unique contents and specialties reflected the particular fortunes of a monastery or cathedral.[19] Library collections were idiosyncratic, depending on factors such as their nearness to trade routes.[20] Some, like Corbie, rose to exceptional prominence, while others languished. Cultural centres with an Irish connection were especially fortunate.[21]

---

authenticity of the work, see Hillgarth, 'The Position of Isidorian Studies', p. 836ff.

[13] All but one of his 13 subjects was a bishop (Codoñer Merino, *El "De viris illustribus" de Ildefonso*).

[14] As Ildefonse explained: *Fertur namque ex antiquitate ueteri, quod potuisse fieri cernitur exemplo temporis novi* (Codoñer Merino, *El "De viris illustribus" de Ildefonso*, p. 112). See also Von Dzialowski, *Isidor und Ildefons*; García Moreno, *Historia*, p. 353.

[15] *Vir egregius adsignans opera uirtutum plus exemplo uiuendi quam calamo scribentis* (Codoñer Merino, *El "De viris illustribus" de Ildefonso*, p. 116).

[16] *Ibidem*, p. 122: *Vir in sacris litteris eruditus, plus uerbis intendens docere quam scriptis*; on Braulio, see *ibidem*, p. 130. The writings of Isidore were given a much fuller treatment (*ibidem*, p. 128).

[17] On the 'spiritual exercise' underlying written traditions, see Hadot, 'Ancient Spiritual Exercises', pp. 126-144.

[18] On the importance of these works for Carolingian librarians, see McKitterick, *The Carolingians*, pp. 200-210, who argues that these works were used as guides for library-building; see also De Ghellinck, *Patristique et moyen âge*, vol, 2, p. 256.

[19] Cf. Ganz, '*Temptabat et scribere*'.

[20] Fontaine, 'Mozarabie hispanique', p. 22.

[21] McKitterick, 'Eighth-century Foundations', pp. 685-686; Bischoff, *Manuscripts and Libraries*, pp. 11-13.

No stable criteria existed for determining who was a Father, nor was it easy to say which books the Fathers had written. Well into the eighth century, the radiance of local centres of learning dominated the concept of literary authority. Moreover, bishops looked to authors of Christian antiquity, not as sources of literary authority, but as one aspect of a greater theme, the provenance of their own spiritual and governmental authority. This authority was expressed above all in the tradition of councils and conciliar law.

The twenty-four Gallican councils held between 314 and 506 adhered to a very spare format which made no room for patristic citations.[22] Similarly, with one notable exception, the Council of Orange in 529,[23] Merovingian assemblies did not look to the Christian authors of late antiquity as a source of law in any systematic way.[24] Leaving aside the Council of Orange, among the fifty-four councils of the Merovingian period there is only one quotation from St. Jerome.[25]

One of the foundational texts of western canon law, the collection of Dionysius Exiguus, displayed a pious attitude toward the ecumenical councils, whose canons it assembled: these were called 'the rules of the saintly Fathers'.[26] Ensuing collections gathered the canons of successive councils, creating a stream of authority that reached from the councils of contemporary times back to the early fourth century, usually arranged in chronological order.[27] This law constituted the 'rule of the Church'.[28]

The conservatism of these councils is often described as a search for authority in the past. In a sense this is true. What should be added is that for

---

[22] Gottlieb, 'Die formalen Bestandteile'.

[23] This council introduced citations of Augustine, Ambrose and Jerome. Council of Orange (529), in De Clercq, *Concilia Galliae*, pp. 53-76. The Merovingian councils may also be consulted in the edition of Maassen (*Concilia aevi merovingici*); Orange (529) is at pp. 44-54. The Council of Orange made use of an Augustinian florilegium made by Prosper of Aquitaine, the *Liber sententiarum ex operibus sancti Augustini delibatarum*. See Cappuyns, 'L'Origine'.

[24] On this question in regard to canon-law collections, see Munier, 'Les sources patristiques'.

[25] The Council of Bordeaux (662-675): *CCL* 148A, pp. 312-313.

[26] Strewe, *Die Canonessammlung* (prologue: *magna est siquidem auctoritas* ... (p. 1); proem of the Council of Chalcedon: *Regulas sanctorum patrum* ... (p. 98). See further on this collection, known as the *Dionysiana*, Gaudemet, *Les sources*, pp. 134-137; on Dionysius: Bardy, *Littérature latine chrétienne*, pp. 194-195.

[27] Mathisen, 'The "Second Council of Arles"', p. 528, and Maassen, *Geschichte*, pp. 640-642. The full records of councils were usually given in compilations which also gathered other important texts, such as papal decretals, secular law, and patristic texts: Fransen, *Les collections canoniques*, pp. 17-19.

[28] Schwartz, 'Die Kanonessammlungen', p. 67.

Gallican and Merovingian bishops there was little distance between themselves and the early Christian past. They embodied that past, especially in council. Councils had a constant concern to revive the 'canons of the ancient fathers'.[29] At Lyon, sometime between 567 and 570, the bishops provided for the necessity of the present time by 'renewing the institutions of the holy fathers'.[30] This renewal of the ancient canons took the form of reissuing laws from previous councils that seemed apropos of the moment, and almost all Gallican and Merovingian councils contained many repeated (and thereby renewed) laws. In considering this phenomenon, what emerges is that episcopal law was not merely a conservative legal system, inching along through precedence, but a basic mode of religious continuity for men who personified apostolic authority in governing society.

By renewing old legislation, bishops affirmed their identity with earlier fathers in a cumulative style of law, linking them to the early Christian past. Patristic literature, also to be found locally, was an annex to the Christian way of life. Gallican and Merovingian bishops relied on the early Christian authors, not as a source for law, but as a source of wisdom supporting the Christian life guided by their legislation, and bolstering their own social and religious position as fathers and heirs of the Apostles.

With the Carolingian usurpation in 751, attitudes toward the Christian past did not change immediately. The cultural centres and libraries of the Merovingian era continued to nourish the Carolingian world, even as their role began to change in a reconfiguration of culture.[31] However, the Frankish church's centre of gravity had moved from the ancient cities of the Mediterranean, to the heartlands of the Carolingian family, giving new prominence to northern cultural centres.[32] The old monastery of St.-Riquier came to house a major library of liturgical and patristic books in the late eighth and ninth centuries.[33] The shift in moral and intellectual equilibrium toward younger churches affected the way bishops looked back upon the Christian past.

By the 780s, Charlemagne had made his court a centre of scholarship, and royal power was turned toward renewing culture, especially liturgical practice.[34] Library collections in Rheims and Cologne became better supplied with patristic works, while Charlemagne also gathered such texts

---

[29] Council of Orléans (549): *CCL* 148A, pp. 148-169 esp. p. 149; Maassen, *Geschichte*, pp. 209-210.

[30] *... pro renouandis sanctorum patrum institutis, quae praesentis temporis necessario fecit opportunitas iterari*: Council of Lyon (567-570), esp. p. 201.

[31] Riché, 'Les centres de culture', p. 300.

[32] McKitterick, *The Frankish Kingdoms*, p. 142.

[33] Dekkers, 'La bibliothèque', pp. 166-175.

[34] Vogel, *La réforme cultuelle*, pp. 214-219.

in his palace library.[35] Scholars now turned to the bibliographic guides mentioned earlier: The *Decretal* of pseudo-Gelasius, the *Illustrious Men* of Jerome-Gennadius, Isidore and Ildefonse.[36] A well-funded monastery such as Murbach could exhibit tremendous ambition in this regard. The monks recorded their desire 'to find many other [books]' they knew of but had never seen.[37]

In the same spirit, new collections of canon law were made.[38] On a trip to Rome in 774, Charlemagne was given a version of the collection of Dionysius Exiguus, known as the *Dionysio-Hadriana*.[39] Under the direction of the palace, ever greater omnibus compilations of law were made.[40] Despite such efforts to gain intellectual control over the mass of old law, however, Carolingian councils did not 'renew the ancient canons' in the manner of Merovingian councils, but broke new ground.

A basic impulse of this cultural movement was to tie the authority of bishops to a centrally defined orthodoxy, thought of as 'Roman'.[41] Kingdom-wide efforts at reform were staged and directed from the royal palace. By the 780s, the king's personal chaplain, in the palace at Aachen, rose to prominence.[42] Because of his unique status and closeness to the king, the chaplain could organise reforms throughout the kingdom,

---

[35] Contreni, 'The Carolingian Renaissance'. On Carolingian library-building, see McKitterick, *Carolingians and the Written Word*, pp. 135-164; Ganz, '*Temptabat et scribere*', p. 26ff.; Wilmart, 'La tradition', pp. 279-292. Gorman ('Wigbod', p. 50), 74-75), argues that Charlemagne gathered patristic texts for the use of scholars such as Wigbod, but strangely enough, goes on to argue that these texts were not really used by anyone. See also *idem*, 'A Carolingian Epitome'; and Bullough, 'Roman Books'.

[36] Dekkers, 'Pour une histoire', p. 68; McKitterick, *Carolingians and the Written Word*, pp. 200-210.

[37] The monks at Murbach relied on the *Institutes* of Cassiodorus as a bibliographic guide. For the phrase *Alios plures invenire desideramus*, see Milde, *Der Biblioteks-katalog*, p. 37.

[38] A collection, to be associated with the Carolingian Renaissance, emulated and expanded the scope of Dionysius Exiguus. Cf. Schieffer, 'Spätantikes Kirchenrecht'.

[39] Wallace-Hadrill, *Frankish Church*, p. 212; Bullough, 'Roman Books', p. 14.

[40] Hen, 'Knowledge of Canon Law', p. 118. One such work, combining the *Dionysio-Hadriana* with the *Vetus Gallica*, reflects the preoccupations of the palace around the year 800. See Mordek, 'Dionysio-Hadriana und Vetus Gallica'.

[41] Carolingian scholars dreamed that their own liturgy was the same as Roman liturgy (Bullough, 'Roman Books', p. 8). The *Epistola generalis* (between 786 and 800) and many other texts urged that chanting be done according to Roman tradition; see Reznikoff, 'Le chant des gaules'.

[42] *Hildebaldus, sacri palatii archiepiscopus* (Preface of the Council of Mainz, 813): *MGH*, *Conc.* 2.1, pp. 258-273. See also Lesne, *La hiérarchie épiscopale*, p. 72; Heuclin, *Hommes de Dieu*, p. 291.

especially by means of grand councils.[43] Around twenty such empire-wide councils were held between 742 and 843.[44]

The Council of Frankfurt in 794 was the first major expression of this cultural shift.[45] What is most striking at first glance is the sheer bulk of the council's activity, far dwarfing any surviving council of the Gallican Merovingian period. The council was a grand expression of royal and episcopal power; and of the highest aims of the Carolingian Empire.[46]

Two great themes were pursued at Frankfurt, both developed in the years leading up to the Council. First, there was an effort to forge an independent position on the question of religous images, which culminated in the *Libri Carolini* of Theodulf.[47] Then came the battle against the Adoptionism of Felix of Urgel and Elipandus of Toledo, which I focus on here. In this conflict, a new method of using patristic authors came into conflict with an older, pre-Carolingian concept of episcopal continuity.

The first canon of the council condemned Adoptionism as an 'abominable heresy', with its perceived claim that Jesus was the Son only by adoption.[48] In support of their position the Carolingians displayed their mastery of patristic literature by gathering portfolios of citations, in a campaign that aimed both to condemn the theology of Toledo, and to forge an Empire-wide orthodoxy. For their part, the clerics of Spain believed that the authenticity of their theology was assured by its presence in the intellectual and liturgical traditions of Toledo.[49]

Elipandus claimed his right to think and teach as a bishop. His position was supported by 'an immense wave of scriptures', a group of writings 'which we teach because we preserve them rightly'.[50] His list of authorities begins in Late Antiquity and then narrows upon the traditions of Spain and

---

[43] Fleckenstein, *Die Hofkapelle*, pp. 48-50.

[44] Hartmann and Jasper, 'Formen und Wege', pp. 43-44. The decisions of Carolingian councils were preserved in a very diffuse manner, ranging from letters of Boniface to the capitularies of the Carolingian kings. See Hartmann, 'La transmission' and Freeman, 'Theodulf of Orléans'.

[45] Council of Frankfurt (794): *MGH Conc.* 2.1, pp. 110-171.

[46] On the combination of royal and episcopal power in the Carolingian state, see Werner, *Naissance de la noblesse*, pp. 360-378.

[47] Freeman, *Opus Caroli regis*. This edition, restoring the original title, replaces the older edition of Bastgen, although the work will probably continue to be known as the *Libri Carolini* (Bastgen, *Libri Carolini, sive Caroli Magni capitulare de imaginibus*).

[48] *Ubi in primordio capitulorum exortum est de impia ac nefanda erese Elipandi Toletane sedis episcopi*: Council of Frankfurt (794), p. 165.

[49] Reynolds, 'The Visigothic Liturgy', p. 920.

[50] *His exemtis, que de inmenso scripturarum pelago*: Council of Frankfurt (794), p. 118.

Toledo in particular: Hilary, Ambrose, Augustine, Jerome, Fulgentius, Isidore, Eugenius, Ildefonse, and Julian.[51] By adhering to the traditions of Toledo, Elipandus claimed, he was a direct heir of all these authors: 'in no way deviating from the vestiges of their decrees'.[52]

The reply of the Frankish bishops to Elipandus presents a stark contrast. The bishops produced a veritable show-piece of patristic learning, drawing on a dozen or more patristic authors.[53] They poured scorn on the traditions of Toledo and Spain as a false and partial tradition: 'If you put these forward as your ancestors, then it is clear what kind of ancestors you have – and hence anyone can see why you were put into the hands of the infidels'.[54] Thus the Moslem invasion of 711 and the conquest of Toledo could be explained as a punishment for the heterodoxy of its bishops, and the peculiar reading-list that went with it.

The attack on Adoptionism marked the appearance among Carolingian bishops of a new conceptualisation of the early Christian past, and of how authentic contact with that past should be made. They had come to view the patristic past as universal and hence above local traditions; as something to be recovered and understood from books, and thus more distant. In 810-812, for example, Charlemagne distributed questionnaires to the archbishops of his kingdom, requiring replies on the theology and practice of baptism.[55] Rather than simple continuity within a see, the catholic tradition was located in a bishop's understanding of patristic authors.[56]

Constant recourse was now made to patristic literature in Carolingian councils, which cannot be explained simply by the greater availability of texts. The councils reflect a new attitude, giving priority to patristic authors over conciliar tradition. Councils held in Aachen in 809, 816 and 817, the

---

51 Council of Frankfurt (794), p. 111.

52 *His premissis sanctorum patrum sententiis assertionibus nostris roboratis, in commune decrevimus ab eorum decretis nullo modo deviare vestigiis*: Council of Frankfurt (794), p. 114.

53 Wallace-Hadrill, *Frankish Church*, pp. 209-211.

54 ... *ut manifestum sit, quales habeatis parentes, et ut notum sit omnibus, unde vos traditi sitis in manus infidelium*: Council of Frankfurt (794), p. 145.

55 Keefe, 'Carolingian Baptismal Expositions'. As Keefe (p. 173) points out, the aim of requiring bishops to write these expositions was to end 'local divergencies in the understanding, teaching, and celebration of baptism'. Such examinations extended beyond baptism; cf. Vogel, *Réforme cultuelle*, pp. 218-223. On the dismay of Amalarius of Trier, see Heuclin, *Hommes de Dieu*, p. 292. Chelini ('Les 'remplois' liturgiques carolingiens', pp. 331-343) has also pointed to this search for authenticity and standardisation in a turn toward patristic literature.

56 For example, a text, probably related to the baptismal examination of 812 betrayed an impulse to handle the symbolism of baptism 'sous le couvert' of ancient authors. Cf, Wilmart, 'Reg. Lat. 69 (fol.116-122)'.

so-called council of Paris in 829, all drew heavily on patristic texts. These clerics stood at the pinnacle of an Empire whose aggressive religious claims mirrored their own universalism.[57] The Fathers, viewed as a distant ideal – a canonical library of lofty books – expressed the aims of these bishops and their king.

The Carolingian vision of the Christian past was a distinctive *mise en scène* in which episcopal and royal aspirations were staged. New awareness of the patristic past involved more than simple devotion to an inheritance. Royal power was mobilised on behalf of this cultural project, because of the implications for imperial and religious unity, and so was the power of bishops, with their special task of reforming the church and kingdom. There is an element of mystery bound up in the process of tradition.[58] While a tradition may establish a union of wills across distances of time and circumstance, it nevertheless transforms what it preserves.[59] Adherence to the past never truly maintains the past. A truer understanding of tradition must therefore take account of the discontinuity of time in respect to written traditions. Carolingian bishops had discovered an archaised patristic tradition, and the distance from it we still feel.

[57] On the power of bishops and their role in governing the Empire, see Kaiser, 'Royauté et pouvoir épiscopal'; *idem*, 'Les évêques de Langres'. On the ideal of imperial unity, see Mohr, 'Die kirchliche Einheitspartei'.

[58] As described by Gadamer (*Praise of Theory*, p. 11), this is a mystery which opens the entire problem of human culture. For a similar large-scale view of tradition, see Shils, *Tradition*, p. 25.

[59] To raise again the case of Augustine's *Confessions*, few fragments of which date before the ninth century: Wilmart ('Manuscrits des "Confessions"', p. 326) remarked, 'Comment s'étonner, après cela, que les écrits de l'antiquité profane se soient mal conservés'?

THE STUDY OF MARTIANUS CAPELLA'S *DE NUPTIIS* IN THE NINTH CENTURY

Mariken Teeuwen

In modern scholarship Martianus Capella's work *De nuptiis Philologiae et Mercurii* has had a bad press. It has been characterised as a monstrous match; an unsuccessful attempt to mix dreary learning with extravagant mythology and satire, written in a kind of Latin that can serve as a textbook example of mannerism. It is full of *hapax legomena,* it employs the most intricate grammatical constructions and rarely used poetical metres, and it is full of Greek vocabulary. Apart from its style, its content is also seriously flawed. Martianus took passages from all kinds of sources, and pasted them together without much skill, which sometimes results in puzzling or even incomprehensible passages.

In the 1970s, however, Fanny Lemoine was the first to attempt to rehabilitate the work and its author in her *Literary Reevaluation,*[1] and since then more modern scholars have pleaded in favour of Martianus. They emphasised the deliberate and well-considered structure of the work, culminating in the recent voluminous monograph of Sabine Grebe, in which the close connection of each book of *De nuptiis* to the others is shown.[2] C.S. Lewis, however, was, compared to his fellow-scholars, very mild when he wrote:

> It is to the same class of mythological allegory that I would assign the work of another writer, if I felt sure that any classification could hold him; for this universe, which has produced the bee-orchid and the giraffe, has produced nothing stranger than Martianus Capella.[3]

Other scholars went so far as to suggest that Martianus must have been senile or 'mentally unwell' when he wrote his work.[4]

Carolingian scholars were not so biased. On the contrary, they saw *De nuptiis* as a *Fundgrube* of ancient knowledge. The mix of work and pleasure, the marriage between story and learning, at which the title of the work

---

[1] Lemoine, *Martianus Capella.*

[2] See, for example, Relihan, *Ancient Menippean Satire*; Grebe, *Martianus Capella.*

[3] Lewis, *The Allegory of Love* (New York, 1963), p. 78.

[4] Lemoine, *Martianus Capella*, pp. 1-3.

already hints, seems to have been very much to their liking. The intricate Latin was seen as a good test case for one's knowledge of the language (on an 'expert level'), and a good opportunity to expand one's vocabulary and grammatical skills. The poems offered new metres. The story-part of the work offered a whole world of Greek myth, and a treatment of the seven liberal arts in a nutshell – these appear as beautiful maidens before Mercury and Philology, and offer them their knowledge as a wedding gift.

There remain today nearly fifty ninth-century manuscripts that bear witness to the popularity of the work.[5] Moreover, most of these are conscientiously annotated, accumulating in their margins information from all kinds of sources and written in a tiny script. In this large number of glosses, several traditions have been recognised. Two corpora of glosses are ascribed with certainty to Carolingian scholars: one to the famous Irish scholar John the Scot, who makes his appearance in Carolingian history as a scholar at the court of Charles the Bald. The other is a scholar of a later generation, Remigius of Auxerre, who lectured on Martianus Capella in Paris at the very end of the ninth century. These corpora of glosses were both edited by Cora Lutz, who made a great contribution to our present state of knowledge on the subject.[6] There is, however, a third, older corpus of glosses, which clearly underlies the other two, be it in two different ways. The commentary of John the Scot, which is considerably smaller and very different in character, is perhaps best understood as a deliberately different opinion on the interpretation of the text, or as an addition to the earlier corpus.[7] Remigius, on the other hand, silently incorporates much of the earliest corpus into his own.

Only part of this oldest commentary has been published in an edition, and many questions have remained unanswered so far. Its author, its place of origin and its date have never been properly established, despite the respectable amount of scholarship produced on the subject, by giants such as Jean Préaux and Claudio Leonardi.[8] In this paper I look at these questions again. Who were responsible for the production of so many Martianus Capella manuscripts in, roughly, the second third of the ninth century? Who

[5] The manuscript tradition has been treated by Préaux, especially in 'Les manuscrits principaux', and by Leonardi ('I codici').

[6] Cf. Lutz, *Johannes Scottus, Annotationes in Marcianum*; *eadem*, *Remigius Autissiodorensis*; *eadem*, 'Martianus Capella'.

[7] See Liebeschütz, 'Zur Geschichte'; Schrimpf, 'Zur Frage der Authentizität'.

[8] See note 5. Others who have made contributions to the state of research in this field are Bishop, 'The corpus Martianus Capella'; Bischoff, *Lorsch im Spiegel seiner Handschriften*, pp. 45, 96-97: second, revised edition, *Die Abtei Lorsch*, pp. 54 and 106-107); McDonough, 'Trier, Bibliothek des Priesterseminars'; Shanzer, 'Review article: Felix Capella'.

established the correct text from the very faulty Late Antique original, and who wrote the oldest layer of glosses? Who were the supposed audience?

These are some of the very basic questions that I encountered on commencing my study of the early medieval reception of this late-ancient source.[9] In fact, because of all these unanswered questions, the making an edition of part of the earliest commentary on Martianus Capella seemed a necessity and took up more time than the study of the actual subject: the Carolingian reception of the ancient *ars musica*. Nevertheless, even after several years of intensive research on a selection of manuscripts of this anonymous commentary, many questions remain unanswered. In this paper, however, I attempt to take the discussion just one step further and tentatively suggest a new candidate for the authorship of these glosses, which opened up this strange text to generations of ninth-, tenth-, eleventh- and even twelfth-century scholars.

Firstly, the state of scholarship on the oldest commentary is briefly sketched. At the end of the nineteenth and beginning of the twentieth century Ludwig Traube and Max Manitius were the first to pay it attention,[10] and attributed it to the Irish bishop Dunchad. In 1944 Cora Lutz edited parts of the commentary under that same name.[11] She based her edition on a single, incomplete source, which was the only one known to her: the manuscript Paris, BnF, lat. 12960 – a late ninth-century miscellany written in the monastery of St.-Pierre in Corbie. The glosses in this manuscript are copied in a running form, that is, without the text of Martianus Capella, and they concern only the last part of the second book (the myth of the marriage of Mercury and Philology), the book on dialectic, and part of the book on rhetoric.

From 1953 onwards Jean Préaux expanded on Lutz's one-item list.[12] He established a list of twelve ninth-century manuscripts with the oldest commentary. Most of these contained both the text of Martianus Capella and the glosses (in interlinear and marginal spaces), and it is generally assumed that this is the original form of the commentary. Moreover, Préaux argued that Dunchad could not have been the author. Instead he pointed to

---

[9] This research project resulted in a dissertation in 2000. A slightly revised version has been published recently: Teeuwen, *Harmony and the Music of the Spheres*.

[10] Traube, 'Computus Helperici', pp. 103-104; Manitius, 'Zur karolingischen Literatur', p. 57; *idem*, *Geschichte der lateinischen Literatur des Mittelalters*, vol. I, pp. 255-256.

[11] Lutz, *Dunchad Glossae in Martianum*.

[12] See, for example, J. Préaux, 'Le commentaire de Martin de Laon sur l'œuvre de Martianus Capella', *Latomus* 12 (1953), pp. 437-459; *idem*, 'Un nouveau manuscrit de Saint-Gall; *idem*, 'Deux manuscrits gantois de Martianus Capella'. The most important article is *idem*, 'Les manuscrits principaux'.

another Irish bishop, Martin of Laon, whose Greek-Latin glossary (Laon, BM, Ms. 444) shows remarkable parallels with the commentary, especially in one long gloss on the names of the nine Muses, their etymological background and allegorical interpretation.[13] Préaux's identification was accepted by many, but not by the scholar who has studied Carolingian Laon in the greatest detail: John Contreni. He argued that Préaux had made a good case for the *possible* identification, but that the actual *evidence* that Martin of Laon had indeed written such a commentary was only weak and circumstantial. According to him there are many more possible candidates: Heiric of Auxerre, Lupus of Ferrières, Haimo of Auxerre, Muridac (the Irish grammarian), John the Scot's student Winibertus, or even Dunchad again.[14] The discussion still finds itself in this stalemate. The one camp is unable to produce stronger arguments, the other is unable to point to just one probable author. The 'Dunchad' or 'Martin of Laon' commentary, in other words, is again 'Anonymous'.

The history of the text of *De nuptiis* is equally uncertain. The text must have survived during the sixth, seventh and eighth centuries in a single manuscript, and is likely to have been copied in the late eighth century in pre-Carolingian *scriptio continua.*[15] When *De nuptiis* gained in popularity in the first half of the ninth century, it was copied in many major and minor centres of scholarship of the Carolingian empire, in particular – so it seems – in the area of northern or north-eastern France: Auxerre, Corbie and Rheims. It is remarkable that among these manuscripts there is abundant evidence of a kind of continuous contact between them, shown, for example, by the additions, corrections and alterations copied in these texts by second and third hands. This has caused modern scholarship to locate the dissemination of the text from one single centre, and Préaux has suggested Corbie.

Because of my primary interest in the *ars musica*, a particular selection of manuscripts that best served my purpose was made. Those manuscripts that contained annotations to the first two mythological books of *De nuptiis*, in which the music of the spheres and numerical speculation are important themes, and manuscripts that contained glossed versions of the ninth and last book of the encyclopedia, *On Music*, were of greatest interest. These criteria led to a selection of four manuscripts of primary relevance, and four of secondary relevance. Their relationship to each other is shown in the table below. A common archetype (or perhaps archetypes) was the basic

[13] Préaux, 'Le commentaire de Martin de Laon'.
[14] Contreni, 'A Note'; *idem*, 'Three Carolingian Texts'.
[15] On the earliest history of the text, see Préaux, 'Les manuscrits principaux', pp. 81-86; Leonardi, 'I codici I', pp. 446-453; and Shanzer, 'Review Article', p. 63.

source for all manuscripts. The four most important manuscripts divide into two pairs. Manuscripts Leiden, UB, Voss.lat.F. 48 and Besançon, BM, Ms. 594 are both dated shortly before 850. They are so closely related that it is difficult to determine which of the two could have been the original and which the copy. The second pair consists of two manuscripts that ended up in Ghent in the eleventh century, though parts of both these manuscripts have been localised in Rheims: Leiden, UB, BPL 88 and Vatican City, BAV, Reg.lat. 1987. Errors in the lay-out of text and glosses show that of these two, Leiden, UB, BPL 88 must have been the original, Vatican City, BAV, Reg.lat. 1987 the copy. Both these manuscripts, furthermore, contain a curious combination of commentary traditions: the Anonymous corpus of glosses is added to books one to eight, but a version of John the Scot's commentary is found in book nine.[16] In Leiden, UB, BPL 88, furthermore, the hand I^2 has been identified as that of a scholar from the circle of John the Scot.[17]

The four manuscripts of secondary relevance[18] are related to the first four in varying degrees of clarity. Oxford, Bodl.Lib., Laud.lat. 118, for example, is related to the Leiden-Besançon pair, but is idiosyncratic in many respects. Leiden, UB, BPL 87 is closest to Leiden, UB, BPL 88, but is much less complete. The manuscripts Paris, BnF, lat. 8671 and Leiden, UB, BPL 36 are related to each other and to the common archetype, but it is unclear through how many intermediary sources. Moreover, the picture represented in the table is very incomplete, since it only matches my particular selection of manuscripts, and several other manuscripts should be added as well to get a more complete picture. Nevertheless, some preliminary observations can be made.

1. Most of the manuscripts stem from the central part of the Carolingian empire: Auxerre, Corbie, Rheims, Northeastern France, and, further to the East, Lorsch have all been suggested on the basis of their palaeographical characteristics. The earliest manuscripts, furthermore, already contain writing or copying errors and errors in layout, which suggest that none of the remaining sources is an original text. The commentary must have existed, therefore, before 850.

[16] A detailed description of codicological and palaeographical aspects of these manuscripts is found in Teeuwen, *Harmony and the Music of the Spheres*, pp. 88-126.

[17] Bishop, 'Autographa of John the Scot', pp. 47-58; Leonardi, 'Glosse eriugeniane'.

[18] These are described in Teeuwen, *Harmony and the Music of the Spheres*, pp. 127-144.

2. Many of these manuscripts have been laid out from their very beginning to contain both text and glosses. Gloss columns are pricked and ruled in their quires, and they are set out with plenty of space between the writing lines, so as to be able to house many interlinear notes.[19] This strongly suggests that there was a more or less authoritative body of glosses, with a fixed form. In other words, these glosses were not random annotations of individual scholars, but copies of an established text. Secondly, it suggests that the text of *De nuptiis* and the Anonymous commentary were originally brought into circulation as a unity. Text and Anonymous commentary seem to be joined from the beginning of Martianus Capella study onwards (that is, the second third of the ninth century) to the period of the dissemination of Remigius's commentary (that is, the beginning of the tenth century) at least.

On the other hand one should note that there are no two manuscripts with the same corpus of glosses. Even those that are closest to each other are different in many respects: lemmata that are annotated in one manuscript are without glosses in the next, and the individual glosses are transmitted with a fair amount of variants.[20] The fluidity of the transmission of these glosses, even when they form a coherent and more or less official corpus seems to be an essential part of their literary status.

3. A puzzling aspect of the Anonymous commentary is its anonymity, even in contemporary sources. Both John the Scot and Remigius of Auxerre often cite the anonymous commentator, but never mention his name. Remigius even goes so far as silently to incorporate the lion's share of the Anonymous commentary into his own. When he cites John the Scot, on the other hand, he often explicitly mentions his name. It seems that Remigius saw the Anonymous commentary as the basis for his own, and as a part of *De nuptiis*.[21]

This anonymity, together with the fact that a relatively large number of copies of *De nuptiis* with the Anonymous commentary appear throughout

---

[19] On the other hand, one should note that this is especially true for the beginnings of most manuscripts, and on that leafing through the manuscripts one usually finds that the spacious layout becomes more and more crowded. Gloss columns are often used for text or are no longer pricked and ruled at all, while the letters grow larger and start filling the larger part of the space between the writing lines.

[20] The frustrating complexity of Latin-Latin gloss material and the trickiness in working with this kind of material is very well (and very wittily) described by Dionisotti, 'On the Nature and Transmission of Latin Glossaries'.

[21] This led Préaux to the conclusion that Remigius was very close to the author of the Anonymous commentary, and that it was perhaps written by one of his teachers: Martin of Laon or Heiric of Auxerre. 'Le commentaire de Martin de Laon'.

the Carolingian empire from roughly the 840s onwards, strongly suggests, to my mind, that text and commentary were brought into circulation on the initiative of a central authority. The court itself, especially under Louis the Pious and Charles the Bald, was a place where monarch and scholars engaged in intellectual conversations about a wide range of theological subjects: the cult of images, predestination, the Trinity and Adoptionism.[22] Secular subjects – which have until now largely been neglected in modern scholarship[23] – seem to have been treated in a similar manner. Classical texts were subjected to textual criticism, corrected and provided with glosses and commentaries; these enriched texts were brought into circulation. The search for the author of the Anonymous commentary could go in that direction: it may very well have been written by a scholar who belonged to the groups of intellectuals collected at the courts of Louis the Pious or Charles the Bald. This would explain both the anonymity of the commentary and the puzzling character of its dissemination.

The character of the glosses also concurs with this hypothesis. One of their primary goals seems to be to establish a correct version of the text. Furthermore, they seem not to be designed for the classroom,[24] but rather to reflect a gathering of as much knowledge as possible. Not only do they guide the reader through grammar and vocabulary, to help him reach a full understanding of the text, they also associatively lead the reader to other Ancient and Late Antique sources. When Martianus touches on the subject of the harmony of the spheres or numerical speculation, Macrobius, Chalcidius and Boethius are invoked; on the subject of mythology, Fulgentius and Servius are quoted; on Greek vocabulary Isidore's etymological explanations are often referred to.

The content of the glosses on music, furthermore, reveals more clues to solve the puzzle of the identity of the author. The glosses on the theme of the harmony of the spheres – a prominent theme, which provoked the writing of many lengthy glosses especially in the first two books of *De nuptiis* – is treated in such a way that a collective authorship is suggested rather than an individual one. The individual glosses are coherent enough, but upon closer inspection they seem not to present a single, unequivocal interpretation of the music of the cosmos. Rather they reflect a discussion on the

[22] There is a large bibliography on this subject, but see among others Freemann, 'Carolingian Orthodoxy'; Ganz, 'The Debate on Predestination'; Chazelle, *The Crucified God in the Carolingian Era*.
[23] An exception is the valuable work of Butzer and Lohrmann, *Science in Western and Eastern Civilization*.
[24] In this respect I do not share the view of Cora Lutz, who has characterised the tone of the Anonymous commentary as didactic and places it in an educational context (*Dunchad*, p. xxvi).

subject, in which different positions are taken and defended against the others. There is a controversy over the direction of the planetary scale: does it go up from the earth to Saturn, or does it go down from the moon to Saturn? Furthermore, there are differing opinions about the actual pitches or tone-steps between the planets.[25] The glosses show a lively debate, in which the widely differing theories of the Late Antique sources which Carolingian scholars had at their disposal are all used, and in which no consensus was achieved.

Hypothetically, the history of the reception of Martianus Capella's *De nuptiis* could be reconstructed as follows. A copy of the text (possibly in *scriptio continua*) was present in the court library of either Louis the Pious or Charles the Bald. The group of intellectuals assembled at their courts took it upon themselves to establish a correct text and to provide it with a commentary. The 'enrichment' of the original text was a collective scholarly venture, and its result was sent to the large cultural centres of the period (Corbie, Auxerre, Laon). From there it was disseminated to the farther corners of the empire. The different scholars who have been suggested as possible candidates for the authorship of the Anonymous commentary (Martin of Laon, Lupus of Ferrières, and the 'Irish intellectual' communities of Laon and Auxerre)[26] could each have made contributions to the project. The existence of such scholarly collaborative ventures instigated by the court has already been proved in the field of theological issues. So far the study of secular texts has not been analysed from this angle, but Michael Bernhard and Calvin Bower, who have prepared an edition of the glosses on Boethius's *De institutione musica*, have reached very similar conclusions.[27] Moreover, it is a fact that Charlemagne's successors, especially Louis the Pious and Charles the Bald, attracted scholars to their courts and consulted them on a wide range of topics. These court 'schools', if they can be given that name, have been described as loosely-organised institutions, with tran-

---

[25] This is analysed in Teeuwen, *Harmony and the Music of the Spheres*, pp. 190-232. It is striking that one of the scholars certainly present at the court of Charles the Bald, namely John the Scot, produced a short treatise on precisely this subject. In this treatise he attacks those who believe in an upward scale from earth to Saturn, and offers a new, unique and controversial opinion instead. See on this Jeauneau, 'Le commentaire érigénien'; Teeuwen, 'Harmonie der hemelsferen'.

[26] Cf. Lutz, *Dunchad*; Préaux, 'Le commentaire de Martin de Laon'; Contreni, 'A Note'; Shanzer, 'Review Article', p. 72 note 60; Ganz, 'Heiric d'Auxerre glossateur', p. 301.

[27] Bernhard and Bower, *Glossa maior in institutionem musicam Boethii*. For an analysis of the relationship between the glosses on music of Martianus Capella and those on Boethius's *De institutione musica*, see Teeuwen, *Harmony and the Music of the Spheres*, pp. 162-183.

sient members, who were often dispersed to monasteries or cathedrals after a stay of several years at the court.[28] It is also known that the libraries of these kings contained many classical and late antique texts, which were explored with great enthusiasm. In the words of John Contreni, the Carolingian leaders instigated 'a well-organised programme, which, given the authority behind it, could not help but achieve its goals'.[29] To my mind, there is enough evidence to assume that Martianus Capella's *De nuptiis* was one of the texts that were, on the initiative of the court, copied in the large monastic scriptoria of Carolingian Europe. On the other hand, the fluidity of the Anonymous corpus of glosses aptly illustrates the 'differences in resources, talents and interests in the cultural landscape', of which Contreni also speaks.[30] The goal of a unitary culture was, also in this respect, illusory, and the plurality of the gloss traditions shows the vitality of learning in the Carolingian era.

[28] See, for example, Sullivan, 'The Context of Cultural Activity'; and Contreni, 'The Pursuit of Knowledge' as well as *idem*, 'The Carolingian Renaissance'. On the courts of Louis the Pious and Charles the Bald, see Godman and Collins, *Charlemagne's Heir*, and Gibson and Nelson, *Charles the Bald.*

[29] Contreni, 'The Carolingian Renaissance', p. 711.

[30] *Ibidem*, p. 712.

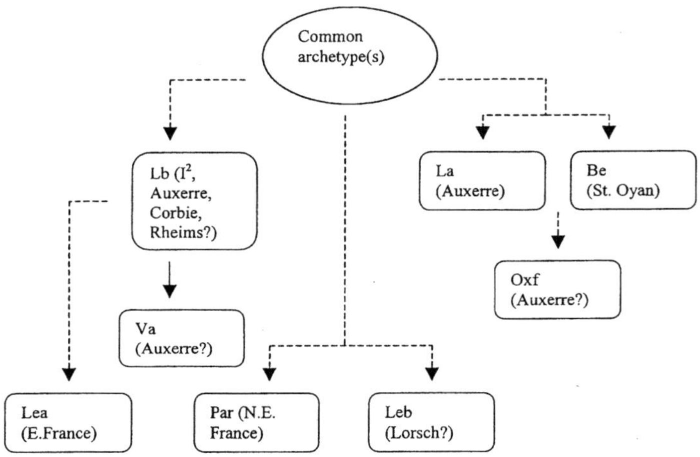

La = Leiden, Universiteitsbibliotheek, Vossianus Latinus Folio 48
Be = Besançon, Bibliothèque Municipale, Ms. 594
Lb = Leiden, Universiteitsbibliotheek, BPL 88
Va = Città del Vaticano, Biblioteca Apostolica Vaticana, Reg.lat. 1987

Lea = Leiden, Universiteitsbibliotheek, BPL 87
Leb = Leiden, Universiteitsbibliotheek, BPL 36
Oxf = Oxford, Bodleian Library, Laud.lat. 118
Par = Paris, Bibliothèque Nationale de France, lat. 8671

BIBLIOGRAPHY

Allard, M., 'Les chrétiens à Bagdād', *Arabica* 9 (1962), pp. 375-388.

Altaner, B., *Patrologie. Leben, Schriften und Lehre der Kirchenväter* (Freiburg, ^{3}1951).

Amelotti, M. and Migliardi Zingale, L., *Le costituzioni giustinianee nei papiri e nelle epigrafi* (Milan, ^{2}1985).

Ando, C., 'Pagan Apologetics and Christian Intolerance in the Ages of Themistius and Augustine', *Journal of Early Christian Studies* 4 (1996), pp. 171-207.

Andresen, J. and Forman, R.K.C., 'Methodological Pluralism in the Study of Religion: How the Study of Consciousness and Mapping Spiritual Experiences Can Reshape Religious Methodology', *Journal of Consciousness Studies* 7 (2000), pp. 7-14.

Andresen, J., ed., *Religion in Mind: Cognitive Perspectives on Religious Belief, Ritual and Experience* (Cambridge, 2001).

Anonymus, 'Excavations in Iraq, 1985-86', *Iraq* 49 (1987), pp. 231-250.

Arnal, W.E., *Jesus and the Village Scribes. Galilean Conflicts and the Setting of Q* (Minneapolis, 2001).

Arnobius, *Adversus nationes libri VII, CSEL* 4, ed. A. Reifferscheid (Vienna, 1885).

Attridge, H.W., 'The Gospel of Truth', in: Attridge, H.W., ed., *Nag Hammadi Codex I (The Jung Codex). Introductions, Texts, Translations, Indices* (Leiden, 1985), pp. 55-122.

Attridge, H.W. and McRae, G.W., 'The Gospel of Truth. Introduction', in: Robinson, J.M., ed., *The Nag Hammadi Library in English* (San Francisco, ^{4}1996), pp. 38-51.

Augustinus, *De utilitate credendi, CSEL* 25, ed. J. Zycha (Prague/Vienna/Leipzig, 1891).

———, *Contra Faustum Manichaeum, CSEL* 25, ed. J. Zycha (Vienna, 1891-1892).

———, *De consensu Evangelistarum libri quattuor, CSEL* 43, ed. F. Weihrich (Vienna, 1904).

———, *The Advantage of Believing*, transl. L. Meagher, pp. 385-442, ed. H. Dressler, *The Fathers of the Church*, vol. 4: *Writings of Saint Augustine* (Washington, 1947).

———, *Confessions*, ed. J.J. O'Donnell, 3 vols. (Oxford, 1992).

Bagnall, R.S., *Egypt in Late Antiquity* (Princeton, 1993).

Bardy, G., *Littérature latine chrétienne* (Paris, 1929).

———, 'Éditions et rééditions d'ouvrages patristiques', *Revue bénédictine* 47 (1935), pp. 356-380.

Bartlett, F., *Remembering. A Study in Experimental and Social Psychology* (Cambridge, 1950).

Bastgen, H., ed., *Libri Carolini, sive Caroli Magni capitulare de imaginibus, MGH Concilia* 2, *Supplementum.*

Baum, W. and Winkler, D.W., *Die apostolische Kirche des Ostens* (Klagenfurt, 2000).

Baumstark, A., *Geschichte der syrischen Literatur* (Bonn, 1922).

Baur, C., 'L'entrée littéraire de saint Jean Chrysostome dans le monde latin', *Revue d'histoire ecclésiastique* 8 (1907), pp. 249-265.

Beaucamp, J., *Le statut de la femme à Byzance (4e-7e siècle)*, vol. 2 (Paris, 1992).

Beaulieu, P.-A., 'The Descendants of Sîn-leqi-unninni', in: Marzahn, J. and Neumann, H., eds., *Assyriologica et Semitica. Festschrift für Joachim Oelsner anläßlich seines 65. Geburtstages am 18. Februar 1997* (Münster, 2000), pp. 1-16.

Bell, H.I., *Greek Papyri in the British Museum* 5 (London, 1917).

———, 'An Egyptian Village in the Age of Justinian', *Journal of Hellenic Studies* 64 (1944), pp. 21-36.

Ben-Amos, D., 'Folklore in the Ancient Near East', in: Freedman, D.N., *et al.*, eds., *Anchor Bible Dictionary* (New York, 1992), vol. 2, pp. 818-828.

Berger, M.S., *Rabbinic Authority* (New York/Oxford, 1998).

Berman, H.J., *Law and Revolution. The Formation of the Western Legal Tradition* (Cambridge, MA, 1983).

Bernhard, M. and Bower, C.M., eds., *Glossa maior in institutionem musicam Boethii* (Munich, 1993, 1994, 1996).

Bidawid, R.J., *Les lettres du patriarche nestorien Timothée I* (Vatican City, 1956).

Bilabel, F., *Berichtigungsliste der griechischen Papyrusurkunden* 2.2 (Heidelberg, 1933).

Bischoff, B., *Lorsch im Spiegel seiner Handschriften* (Munich, 1974); second, revised edition: *Die Abtei Lorsch im Spiegel ihrer Handschriften* (Lorsch, 1989).

———, *Manuscripts and Libraries in the Age of Charlemagne*, transl. M. Gorman (Cambridge, 1994).

Bishop, T.A.M., 'The corpus Martianus Capella. Welsh and Continental Tradition, Drawn upon by the Revived Latin Learning of Tenth-Century England, United in Corpus Christi College 153 (Martianus Capella)', *Transactions of the Cambridge Bibliographical Society* 4 (Cambridge, 1949), pp. 257-275.

———, 'Autographa of John the Scot', in: *Jean Scot Erigène et l'histoire de la philosophie* (Paris, 1977), pp. 47-58.

Black, J.A. and Wiseman, D.J., *Literary Texts from the Temple of Nabû* (Oxford, 1996).

Boethius, *De consolatione philosophiae. Opuscula theologica*, ed. C. Moreschini (Munich, 2000).

Boiy, T., *Laatachaemenidisch en hellenistisch Babylon. Portret van een Mesopotamische stad in een cultureel spanningsveld* (Louvain, 2000) [*unpublished dissertation*].

Bongenaar, A.C.V.M., *The Neo-Babylonian Ebabbar Temple at Sippar: Its Administration and its Prosopography* (Istanbul, 1997).

———, 'Private archives in Neo-Babylonian Sippar and their Institutional Connections', in: Bongenaar, A.C.V.M., ed., *Interdependency of Institutions and Private Entrepreneurs (MOS Studies 2), Proceedings of the Second MOS Symposium (Leiden, 1998)* (Istanbul, 2000), pp. 73-94.

Bonnet, M., 'Ex actis Andreae', in: *Acta Apostolorum Apocrypha* II.1 (Darmstadt, 1959 [Leipzig, 1898]), pp. 38-45.

Bos, A.P., *Providentia divina. The Theme of Divine Pronoia in Plato and Aristotle* (Amsterdam, 1976).

———, *Cosmic and Meta-Cosmic Theology in Aristotle's Lost Dialogues* (Leiden, 1989).

Bottéro, J., 'Symptômes, signes, écritures en Mesopotamie ancienne', in: Vernant, J.-P., ed., *Divination et Rationalité* (Paris, 1974), pp. 70-197.

———, 'Religion and Reasoning in Mesopotamia', in: Bottéro, J., Herrenschmidt, C. and Vernant, J.-P., *Ancestor of the West* (Chicago, 2000), pp. 1-66.

Bovon, F., 'The Words of Life in the Acts of Andrew', in: Bremmer, J.N., ed., *The Apocryphal Acts of Andrew* (Louvain, 2000), pp. 81-95.

Bowersock, G.W., *Hellenism in Late Antiquity* (Ann Arbor, 1990).

Bowker, J., *The Targums and Rabbinic Literature: An Introduction to Jewish Interpretations of Scripture* (Cambridge/New York, 1967).

Boyancé, P., 'Sur les mystères d'Éleusis', *Revue des Études grecques* 75 (1962), pp. 460-473.

———, 'Le Dieu très haut chez Philon', in: Lévy, P. and Wolff, E., eds., *Mélanges d'histoire des religions offerts à H.-C. Puech* (Paris, 1974), pp. 139-149.

Braun, O., ed. and transl., *Timothei patriarchae I epistulae*, I, *CSCO*, 74, 75, *Scriptores Syri*, 30, 31 (Paris, 1914, 1915).

Bretone, M., *Storia del diritto romano* (Rome/Bari, [4]1991)

Britton, J. and Walker, C.B.F., 'A 4th Century Babylonian Model for Venus: B.M. 33552', *Centaurus* 34 (1991), pp. 97-118.

Britton, J., 'Lunar Anomaly in Babylonian Astronomy', in: Swerdlow, N.M. ed., *Ancient Astronomy and Celestial Divination* (Cambridge, MA, 1999), pp. 187-254.

BRM 1 = Caly, A.T., *Babylonian Business Transcations of the First Millennium B.C.* (New York, 1912).

Brock, S., 'Syriac Historical Writing: A Survey of the Main Sources', *Journal of the Iraqi Academy (Syriac Corporation)* 5 (1979/80), pp. 1-30; repr. in: Brock, *Studies in Syriac Christianity*, vol. I.

———, 'From Antagonism to Assimilation: Syriac Attitudes to Greek Learning', in: Garsoïan, N., Mathews, T. and Thompson, R., eds., *East of Byzantium: Syria and Armenia in the Formative Period (Dumbarton Oaks Symposium, 1980)* (Washington, 1982), pp. 17-34; repr. in: Brock, *Syriac Perspectives*, vol. V.

———, *Syriac Perspectives on Late Antiquity* (London, 1984).

———, *Studies in Syriac Christianity* (Aldershot, 1992).

———, 'Two Letters of the Patriarch Timothy from the Late Eighth Century on Translations from Greek', *Arabic Sciences and Philosophy* 9 (1999), pp. 233-246.

Brody, R. and Herr, M.D., eds., *Ephraim E. Urbach, Collected Writings in Jewish* (Jerusalem, 1999).

Brooks, E.W., ed. and transl., *Historia ecclesiastica Zachariae rhetori vulgo adscripta*, I, II, *CSCO*, 83-84, 87-88, *Scriptores Syri*, 38-39, 41-42 (Louvain, 1919, 1921, 1924).

Brown, D.R., *Mesopotamian Planetary Astronomy-Astrology* (Groningen, 2000).

———, 'The Cuneiform Conception of Celestial Space and Time', *Cambridge Archaeological Journal* 10 (2000), pp. 103-121.

———, 'Astronomy-Astrology in Mesopotamia', *Bibliotheca Orientalis* 58 (2001), pp. 41-59.

———, 'Misinformation on Mesopotamian Exact Sciences', in: Abusch, T., Noyes, C., Hallo, W.W. and Winter, I., eds., *Proceedings of the XLV*[e] *Rencontre Assyriologique Internationale held at Harvard and Yale, USA, July 1998. Historiography in the Cuneiform World* (Bethesda, 2001), pp. 79-89.

———, 'Astral Divination and Related Matters', in: Cullen, C., ed., *Festschrift for Ho Peng Yoke*). Forthcoming.

———, 'Parallel Approaches to the Stars in Mesopotamia and Greece'. Forthcoming.

Brown, D.R. and Linssen, M., 'BM 134701 = 1965-10-14,1 and the Hellenistic Period Eclipse Ritual from Uruk', *Revue d'Assyriologie* 91 (1997), pp. 147-166.

Brown, P., *The Body and Society. Men, Women and Sexual Renunciation in Early Christianity* (London/Boston, 1988).

Browne, T., *Pseudodoxia epidemica* (London, 1646, [6]1672).

Browning, R., 'Education in the Roman Empire', in: Cameron, Averil, Ward-Perkins, B. and Whitby, M., eds., *The Cambridge Ancient History*, vol. 14: *Late Antiquity: Empire and Successors, AD 425-600* (Cambridge, 2000), pp. 855-883.

Bucking, S., 'Christian Educational Texts from Egypt: A Preliminary Inventory', in: Kramer, B., ed., *Akten des 21. Internationalen Papyrologenkongresses, Berlin 1995* (Stuttgart, 1997), pp. 132-138.

Budge, E.A.W., ed. and transl., *Syrian Anatomy, Pathology and Therapeutics or 'The Book of Medicines'*, 2 vols. (London, 1913).

Bullough, D., 'Roman Books and Carolingian *renovatio*', in: Bullough, D., ed., *Carolingian Renewal: Sources and Heritage* (Manchester, 1991), pp. 1-38.

Burstein, S.M., *The Babyloniaca of Berossus* (Malibu, 1978).

Butzer, P.L. and Lohrmann, D., eds., *Science in Western and Eastern Civilization in Carolingian Times* (Basel/Boston/Berlin, 1993).

Cameron, Alan, 'The Poet, the Bishop, and the Harlot', *Greek, Roman and Byzantine Studies* 41 (2000), pp. 175-188.

Cameron, Averil, 'Wandering Poets: A Literary Movement in Byzantine Egypt', *Historia* 14 (1965), pp. 470-509.

———, *Literature and Society in the Early Byzantine World* (London, 1985) Chapter I ('Wandering Poets: a Literary Movement in Byzantine Egypt', 1965).

———, 'The Empress and the Poet: Paganism and Politics at the Court of Theodosius II', *Yale Classical Studies* 27 (1982), pp. 217-289.

———, *Procopius and the Sixth Century* (London, 1985).

———, *The Mediterranean World in Late Antiquity* (London, 1993).

Cappuyns, D.M., 'L'Origine des 'Capitula' d'Orange 529', *Recherches de théologie ancienne et médievale* 6 (1934), pp. 121-142.

Cavigneaux, A., *Textes scolaires du Temple de Nabû ša Hare*, vol. I (Baghdad, 1981).

CCL: Corpus Christianorum Latinorum

Ceresa-Gastaldo, A., ed., *Gerolamo, Gli uomini illustri* (Florence, 1988).

Chabot, J.B., ed. and transl., *Le livre de la chasteté composé pas Jésusdenah, évêque de Baçrah* (Rome, 1896).
Chazelle, C., *The Crucified God in the Carolingian Era* (Cambridge, 2000).
Chediath, G., *The Christology of Mar Babai the Great* (Kottayam/Paderborn, 1982).
Chelini, J., 'Les 'remplois' liturgiques carolingiens', *Settimane di Studio del Centro Italiano di studi sull'alto medioevo* 46 (1999), pp. 321-357.
Chernick, M., '"Turn it and turn it again": Culture and Talmud Interpretation', *Exemplaria* XII/1 (2000), pp. 63-104.
Christes, J., 'Enkyklios Paideia', *Der Neue Pauly*, vol. 3 (Stuttgart-Weimar 1997), 1037-1039.
Cicero, *De goden*, transl. V. Hunink; intr. J. den Boeft (Amsterdam, 1993).
———, *De natura deorum. Academica*, ed. and transl. H. Rackham (Cambridge, MA, 1972).
———, *De officiis* (Latin and English by W. Miller (Cambridge, MA, 1913; repr. 1975).
———, *De oratore Books I, II* (Latin and English), transl. E.W. Sutton (Cambridge, MA, 1942; repr. 1976).
Cirillo, L., 'L'Uomo interiore negli Atti apocrifi di Andrea e il 'nous' nella visione di Mani secondo Baraies (CMC 17.7ss)', in: Cirillo, L. and Tangerloo, A. eds., *Manichaean Studies* III. *Manicheismo e oriente cristiano. Atti del Terzo Congresso Internazionale di Studi* (Louvain/Naples, 1997), pp. 11-21.
Cizek, E., *Histoire et historiens à Rome dans l'antiquité* (Lyon, 1995).
Clarysse, W., ed., *Leuven Database of Ancient Books* (*LDAB*) (Louvain, 1998, [2]2001).
Claus, A., *'Ο σχολαστικός* (Cologne, 1965).
Codoñer Merino, C., ed., *El "De viris illustribus" de Isidoro de Sevilla* (Salamanca, 1964).
Cole, M. *et al.*, *The Cultural Context of Learning and Thinking* (New York, 1971).
Congar, Y., *La Tradition et les traditions*, 2 vols. (Paris, 1960-1963).
Conrad, L.I., Neve, M., Nutton, V., Porter, R. and Wear, A., *The Western Medical Tradition 800 BC to AD 1800* (Cambridge, 1995).
Contreni, J.J., 'A Note on the Attribution of a Martianus Capella Commentary to Martinus Laudunensis (Martinus Scottus)', in: Kristeller, Cranz *et al.*, *Catalogus translationum et commentariorum*, vol. III, pp. 451-452.
———, 'Three Carolingian Texts attributed to Laon: Reconsiderations', *Studi Medievali Serie Terza* 17 (1976), pp. 797-813.
———, 'The Carolingian Renaissance', in: Treadgold, W., ed., *Renaissances Before the Renaissance: Cultural Revivals of Late Antiquity and the Middle Ages* (Stanford, 1984), pp. 59-74.
Contreni, J.L., 'The Carolingian Renaissance: Education and Literary Culture', in: *The New Cambridge Medieval History*, vol. II, *c. 700–c. 900* (Cambridge, 1995), pp. 709-757.
———, 'The Pursuit of Knowledge in Carolingian Europe', in: Sullivan, R.E., ed., *The Gentle Voices of Teachers. Aspects of Learning in the Caro-lingian Age* (Ohio UP, 1995), pp. 106-141.
Cribiore, R., *Writing, Teachers, and Students in Graeco-Roman Egypt* (Atlanta, 1996).

———, *Gymnastics of the Mind: Greek Education in Hellenistic and Roman Egypt* (Princeton, 2001).

Croissant, J., *Aristote et les mystères* (Paris, 1932).

Crossan, J.D., *The Birth of Christianity. Discovering what Happened in the Years Immediately After the Execution of Jesus* (San Francisco, 1998).

Cryer, F.H., *Divination in Ancient Israel and its Near Eastern Environment – a Socio-Historical Investigation* (Sheffield, 1994).

CSCO: *Corpus Scriptorum Christianorum Orientalium.*

CSEL: *Corpus Scriptorum Ecclesiasticorum Latinorum.*

CT 49= Kennedy, D.A., *Late-Babylonian economic Texts* (London, 1968).

Culmann, O., 'Infancy Gospels', in: Schneemelcher, W. and McL. Wilson, R., eds., *New Testament Apocrypha*, vol. 1 (Cambridge/Louisville, [2]1991), pp. 414-469.

Czapla, B., ed., *Gennadius als Litterarhistoriker. Eine quellenkritische Untersuchung der Schrift des Gennadius von Marseille 'De viris illustribus'* (Münster, 1898).

De Clercq, C., ed., *Concilia Galliae A.511- A.695*, *CCL* 148A (Turnhout, 1963).

De Ghellinck, J., *Patristique et moyen âge: Études d'histoire littéraire et doctrinale*, vol. 2: *Introduction et compléments à l'étude de la patristique* (Gembloux, 1947).

De Lange, N., 'The Revival of the Hebrew Language in the Third Century', *Jewish Studies Quarterly* 3 (1996), pp. 342-358.

De Zulueta, F., *The Institutes of Gaius*, vol. I (Oxford 1946).

Degen, R., 'Ein Corpus Medicorum Syriacorum', *Medizinhistorisches Journal* 7 (1972), pp. 114-122.

———, 'Galen im Syrischen: Eine Übersicht über die syrische Überlieferung der Werke Galens', in: Nutton, V., ed., *Galen: Problems and Prospects* (London, 1981), pp. 131-166.

Dekkers, E., 'La bibliothèque de Saint-Riquier au moyen âge, viie-xiie siècle', *Bulletin des antiquaires de Picardie* (1956), pp. 157-197.

———, E., 'Pour une histoire de la bibliographie chrétienne dans l'Antiquité et au Moyen Age', in: *Aevum inter utrumque. Mélanges offerts à Gabriel Sanders* (The Hague, 1991), pp. 53-65.

DePalma Digester, E., *The Making of a Christian Empire. Lactantius & Rome* (Ithaca/London, 2000).

Deuse, W., *Untersuchungen zur mittelplatonischen und neuplatonischen Seelenlehre* (Mainz, 1983).

Dietrich, A., *Medicinalia Arabica. Studien über arabische medizinische Handschriften in türkischen und syrischen Bibliotheken* (Göttingen, 1966).

Digesta Iustiniani Augusti. Editio altera, ed. Th. Mommsen, 2 vols. (Berlin, 1962).

Dillon, J., *Alcinous. The Handbook of Platonism* (Oxford, 1993).

Dionisotti, A.C.,'On the Nature and Transmission of Latin Glossaries', in: Hamesse, J., ed., *Les manuscrits de lexiques et glossaires de l'antiquité au moyen âge* (Louvain-la-Neuve, 1996), pp. 202-252.

Dodds, E.R., *Proclus. The Elements of Theology. A Revised Text with Translation, Introduction and Commentary* (Oxford, [2]1963).

Dols, M.W., 'The Origins of the Islamic Hospital: Myth and Reality', *Bulletin of the History of Medicine* 61 (1987), pp. 367-390.

———, 'Syriac into Arabic: The Transmission of Greek Medicine', *Aram* 1 (1989), pp. 45-52.

Donini, P.L., *Tre studi sull'Aristotelismo nell II secolo D. C.* (Torino, 1974).

Donne, John, *Selected Poetry*, ed. M. Bewley (New York/Toronto/London, 1966).

Dormeyer, D., *The New Testament among the Writings of Antiquity*, transl. R. Kossov (Sheffield, 1998).

Dörrie, H., 'Zum Ursprung der neuplatonischen Hypostasenlehre', *Platonica Minora* (Munich, 1976), pp. 286-296.

Drews, R., 'The Babylonian Chronicles and Berossus', *Iraq* 37 (1975), pp. 39-55.

Durand, J.-M., *Archives Royales de Mari 26/1, Archives épistolaires de Mari 1* (Paris, 1988).

Edwards, M., Goodman, M. and Price, S., eds., *Apologetics in the Roman Empire* (Oxford, 1999).

Elgood, C., *A Medical History of Persia and the Eastern Caliphate* (Cambridge, 1951, repr. Amsterdam, 1979).

Elm, S., *'Virgins of God'. The Making of Asceticism in Late Antiquity* (Oxford, 1994).

Encyclopedia of Philosophy, ed. P. Edwards, 8 vols. (New York, 1967).

Erbetta, M., 'Gli Atti di Andrea', *Gli Apocrifi del Nuovo Testamento*, vol. II (Turin, 1966), pp. 394-449.

Eutropius, *Breviarium ab urbe condita*, ed. F.L. Müller (Stuttgart, 1995).

Evans, J.A.S., *The Age of Justinian* (London/New York, 1996).

Falkenstein, A., *Literarische Keilschrifttexte aus Uruk* (Berlin, 1931).

Farber, W., 'Neues aus Uruk: Zur Bibliothek des Iqiša', *Welt des Orients* 18 (1987), pp. 26-42.

Feder, A., *Studien zum Schriftstellerkatalog des hl. Hieronymus* (Freiburg, 1927).

Feldman, L.H., *Jew and Gentile in the Ancient World. Attitudes and Interactions from Alexander to Justinian* (Princeton, 1993).

———, 'Reflections on Rutgers' "Attitudes to Judaism in the Greco-Roman Period"', *Jewish Quarterly Review* 86 (1995), pp. 153-170.

———, 'Some Observations on Rabbinic Reaction to Roman Rule in Third Century Palestine', *Studies in Hellenistic Judaism* (Leiden, 1996), pp. 438-483.

———, 'Torah and Secular Culture: Challenge and Response in the Hellenistic Period', *Studies in Hellenistic Judaism* (Leiden, 1996), pp. 487-503.

Festugière, A.J., *La Révélation d'Hermes Trimégiste* IV (Paris, 1954).

Fichtenau, H., *The Carolingian Empire: The Age of Charlemagne*, transl. P. Munz (1957; repr. New York, 1964).

Fiey, J.M., *Assyrie chrétienne*, vol. III (Beirut, 1968).

Finkel, I.L., 'On Late Babylonian Medical Training', in: George, A.R. and Finkel, I.L., eds., *Wisdom, Gods and Literature. Studies in Assyriology in Honour of W.G. Lambert* (Winona Lake, 2000), pp. 137-223.

Firmicus Maternus, *De errore profanarum religionum*, *CSEL* 2, ed. C. Halm (Vienna, 1867).

Fitzgerald, J.T., 'Virtue/Vice Lists', in: Freedman, D.N., *et al.*, eds., *The Anchor Bible* Dictionary, vol. 6 (New York, 1992), pp. 857-859.

Fleckenstein, J., *Die Hofkapelle der deutschen Könige 1: Grundlegung. Die karolingische Hofkapelle* (Stuttgart, 1959).

Fontaine, J., 'Mozarabie hispanique et monde carolingien: Les echanges culturels entre la France et l'Espagne du VIIIe au Xe siècle', *Anuario de Estudios medievales* 13 (1983), pp. 17-46.

Fotinis, A., *The* De Anima *of Alexander of Aphrodisias. A Translation and Commentary* (Washington, 1979).

Fournet, J.-L., 'Appendice sur le nom d''Αφροδίτης κώμη', *Revue des Études Grecques* 105 (1992), pp. 235-236.

———, 'Une éthopée de Caïn dans le Codex des Visions de la Fondation Bodmer', *Zeitschrift für Papyrologie und Epigraphik* 92 (1992), pp. 253-266.

———, 'À propos de *SB* XIV 11856 ou Quand la poésie rencontre le document', *Bulletin de l'Institut Français d'Archéologie Orientale* 93 (1993), pp. 223-235.

———, 'Un nouvel épithalame de Dioscore d'Aphrodité adressé à un gouverneur civil de Thébaïde', *Antiquité Tardive* 6 (1998), pp. 65-82.

———, *Hellénisme dans l'Égypte du VIe siècle* (Paris, 1999).

———, 'Le système des intermédiaires dans les reçus fiscaux byzantins et ses implications chronologiques sur le dossier de Dioscore d'Aphrodité', *Archiv für Papyrusforschung* 46 (2000), pp. 233-247.

———, 'Entre document et littérature: la pétition dans l'antiquité tardive', in: *Actes du XXe Congrès International des Études Byzantines, Paris 2001*. Forthcoming.

Fransen, G., *Les collections canoniques* (Turnhout, 1973).

Freemann, A., 'Carolingian Orthodoxy and the Fate of the *Libri Carolini*', *Viator* 16 (1985), pp. 65-108.

———, 'Theodulf of Orléans: A Visigoth at Charlemagne's Court', in: Fon-taine, J. and Pellistrandi, C., eds., *L'Europe héritière de l'Espagne wisi-gothique* (Madrid, 1992), pp. 185-194.

———, ed., *Opus Caroli regis contra synodum (Libri Carolini), MGH Con-cilia* 2, Suppl. 1 (Hanover, 1998).

Frend, W.H.C., *The Rise of the Monophysite Movement* (London, 1972).

Frier, B., *The Rise of the Roman Jurists* (Princeton, 1985).

Frye, R.N., *The Golden Age of Persia: The Arabs in the East* (London, 2000).

Gadamer, H.G., 'Culture and the Word', in: Gadamer, H.-G., *Praise of Theory: Speeches and Essays*, transl. C. Dawson (New Haven, 1998), pp. 1-15.

Gagos, T. and Bagnall, R.S., eds., *Essays and Texts in Honor of J. David Thomas* (*s.l.*, 2001).

Gagos, T. and Van Minnen, P., *Settling a Dispute: Toward a Legal Anthropology of Late Antique Egypt* (Ann Arbor, 1994).

Gamble, H.I., *Books and Readers in the Early Church. A History of Early Christian Texts* (New Haven/London, 1995).

Ganz, D., 'The Debate on Predestination', in: Gibson, M.T. and Nelson, J.L., eds., *Charles the Bald: Court and Kingdom* (Aldershot, ^{2}1990), pp. 281-302.

———, 'Heiric d'Auxerre glossateur du Liber glossarum', in: Iogna-Prat, D., Jeudy, C. and Lobrichon, G., eds., *L'école carolingienne d'Auxerre. De Murethach à Rémi 830-908* (Paris, 1991), p. 301.

———, '*Temptabat et scribere*: Vom Schreiben in der Karolingerzeit', in: Schieffer, R., ed., *Schriftkultur und Reichsverwaltung unter den Karolingern. Referate des Kolloquiums der Nordrhein-Westfälischen Akademie der Wissenschaften am 17./18. Februar 1994 in Bonn* (Opladen, 1996), pp.

13-33.
García Moreno, L.A., *Historia de España visigoda* (Madrid, 1989).
Gätje, H., *Grundriß der Arabischen Philologie, Band II: Literaturwissenschaft* (Wiesbaden, 1987).
Gauchet, M., *Désenchantement du monde: une histoire politique de la religion* (Paris, 1985).
Gaudemet, J., *Les sources du droit de l'Église du IIe au VIIe siècle* (Paris, 1985).
Gesche, P.D., *Schulunterricht in Babylonien im ersten Jahrtausend v. Chr.* (Münster, 2001).
Gibson, M.T. and Nelson, J.L., eds., *Charles the Bold: Court and Kingdom* (Aldershot, ²1990).
Gignoux, P., 'Le traité syriaque anonyme sur les médications', in: Levenant, R., ed., *Symposium Syriacum VII* (Rome, 1998), pp. 725-733.
Godman, P. and Collins, R., eds., *Charlemagne's Heir: New Perspectives on the Reign of Louis the Pious (814-840)* (Oxford, 1990).
Golding, Arthur, transl., *The Excellent and Pleasant Worke Collectanea rerum memorabilium of Gaius Julius Solinus* (1587) facsimile, intr. G. Kish (1587) (Gainesville, Florida, 1955).
Golega, J., *Der homerische Psalter* (Ettal, 1960).
Gorman, M., 'A Carolingian Epitome of St. Augustine's *De Genesi ad Litteram*', *Revue des études augustiniennes* 29 (1983), pp. 137-144.
———, 'Wigbod and Biblical Studies under Charlemagne', *Revue bénédictine* 107 (1997), pp. 40-76.
Gottlieb, G., 'Die formalen Bestandteile in der Überlieferung der gallischen Konzilien des 4. und 5. Jahrhunderts', *Annuarium Historiae Conciliorum* 16 (1984), pp. 254-263.
Grant, M., *Galen on Food and Diet* (London/New York, 2000).
Grebe, S., *Martianus Capella, 'De nuptiis Philologiae et Mercurii': Darstellung der Sieben Freien Künsten und ihre Beziehungen zueinander* (Stuttgart, 1999).
Grenfell, B.P. and Hunt, A.S., eds., *The Oxyrhynchus Papyri*, vol. I (London, 1898).
Groome, D., *et al.*, *An Introduction to Cognitive Psychology. Processes and Disorders* (London/New York, 1999).
Guelich, R.A., *Mark 1:8-26* (Dallas, 1989).
Gutas, D., *Greek Thought, Arabic Culture: The Graeco-Arabic Translation Movement in Baghdad and Early 'Abbāsid Society (2nd-4th/8th-10th centuries)* (London/New York, 1998).
Guyotjeannin, O., Pycke, J. and Tock, B.-M., *Diplomatique médiévale* (Turnhout, 1993).
Hadot, P., 'Ancient Spiritual Exercises and "Christian Philosophy"', in: *Philosophy as a Way of Life*, transl. M. Chase (Oxford, 1995), pp. 126-144.
Hagedorn, D. and Kramer, B., *Griechische Papyrusurkunden der Hamburger Staats- und Universitätsbibliothek* 3 (Bonn, 1984).
Hamilton, F.J. and Brooks, E.W., *The Syriac Chronicle Known as that of Zachariah of Mitylene* (London, 1899).
Hardy, E.R., *The Large Estates of Byzantine Egypt* (New York, 1931).
Hartmann, W. and Jasper, D., 'Formen und Wege mittelalterlicher Konzilsüberlieferungen', in: *Mittelalterliche Textüberlieferungen und ihre kritische Aufarbeitung: Beiträge der Monumenta Germaniae Historica zum 31. deut-*

schen Historikertag, Mannheim 1976 (Munich, 1976), pp. 42-51.

Hartmann, W., 'La transmission et l'influence du droit synodal carolingien', *Revue historique de droit français et étranger*, 4th ser. 63 (1985), pp. 483-497.

Hasitzka, M.R.M., *Neue Texte und Dokumentation zum Koptisch-Unterricht* (Vienna, 1990).

Hau, F.R., 'Gondeschapur – eine Medizinschule aus dem 6. Jahrhundert n. Chr.', *Gesnerus* 36 (1979), pp. 98-115.

Heeßel, N., *Babylonisch-assyrische Diagnostik* (Münster, 2000).

Heitsch, E., *Die Griechischen Dichterfragmente der römischen Kaiserzeit* (Göttingen, 1964).

Hen, Y., 'Knowledge of Canon Law among Rural Priests: The Evidence of Two Carolingian Manuscripts from around 800', *The Journal of Theological Studies* 50 (1999), pp. 117-134.

Hennecke, E., 'Andreasakten', *Handbuch zu den neutestamentliche Apokryphen* (Tübingen, 1904), p. 544-562.

Heuclin, J., *Hommes de Dieu et fonctionnaires du roi en Gaule du Nord du Ve au IXe siècle (348-817)* (Villeneuve-d'Ascq, 1998).

Hezser, C., *Jewish Literacy in Roman Palestine* (Tübingen, 2001).

Hickey, T.M., 'A Fragment of a Letter from a Bishop to a Scholastikos', *Zeitschrift für Papyrologie und Epigraphik* 110 (1996), pp. 127-134.

Hilhorst, A., 'The Cleansing of the Temple (John 2, 13-25) in Juvencus and Nonnus', in: Den Boeft, J. and Hilhorst, A., eds., *Early Christian Poetry* (Leiden, 1993), pp. 61-76.

Hillgarth, J.N., 'The Position of Isidorian Studies: A Critical Review of the Literature, 1936-1975', *Studi Medievali*, 3d. ser. 24.2 (1983), pp. 817-905.

Honoré, T., *Tribonian* (London, 1978).

———, 'Ausonius and Vulgar Law', *Iura* 35 (1984-1987), pp. 75-85.

Hopkins, K., *A World Full of Gods: pagans, Jews and Christians in the Roman Empire* (London, 1999).

Hopkinson, N., ed., *Studies in the Dionysiaca of Nonnus* (Cambridge, 1994).

Horowitz, W., *Mesopotamian Cosmic Geography* (Winona Lake, 1998).

Horsley, G.H.R., *New Documents Illustrating Early Christianity* (North Ryde, 1989).

Horsley, R.A., *Archeology, History, and Society in Galilee, The Social Context of Jesus and the Rabbis* (Valley Forge, 1996).

Hugonnard-Roche, H., 'Note sur Sergius de Reš'ainā, traducteur du grec en syriaque et commentateur d'Aristote', in: Endress, G. and Kruk, R., eds., *The Ancient Tradition in Christian and Islamic Hellenism. Studies on the Transmission of Greek Philosophy and Sciences Dedicated to H.J. Drossaart Lulofs on his Ninetieth Birthday* (Leiden, 1997), pp. 121-143.

Hummel, K., 'Die Anfänge der iranischen Hochschule Gundischapur in der Spätantike', *Tübinger Forschungen* 9 (1963), pp. 1-4.

Hunger, H., *Babylonische und assyrische Kolophone* (Neukirchen-Vluyn, 1968).

Hunger, H., 'Holztafel', in: *Reallexikon der Assyriologie*, vol. 4 (Berlin/New York, 1972-1975), pp. 458-459.

———, *Spätbabylonische Texte aus Uruk*, vol. 1 (Berlin, 1976).

———, *Astrological Reports to Assyrian Kings* (Helsinki, 1992).

———, *Astronomical Diaries and Related Texts Volume V – Lunar and Planetary Texts* (Vienna, 2001).

Hunger, H. and Pingree, D., *MUL.APIN – An Astronomical Compendium in Cuneiform, Archiv für Orientforschung Beiheft* 24 (Horn, 1988).

Hunt, A.S. and Edgar, C.C., *Select Papyri* 1 (London/Cambridge, MA, 1932).

Hurst, A. and Rudhardt, J., *"Codex des Visions". Poèmes divers,* Papyri Bodmer XXX-XXXVII (Munich, 1999).

Hurst, A., Reverdin, O. and Rudhardt, J., *Vision de Dorothéos*, Papyrus Bodmer XXIX (Cologny/Geneva, 1984).

Hyskell, I.D., *A Study of the Latinity of Solinus* (Chicago, 1925).

Jacobs, H.H., 'Die große Zeit der Glossatoren', *Zeitschrift der Savigny-Stiftung Rom. Abt.* 116 (2000), pp. 222-258.

Jacoby, F., *Die Fragmente der griechischen Historiker*, vol. 3. *Die Geschichte von Städten und Volker (Horographie und Ethnographie), C. Autoren über einzelne Länder, (Nr. 608a-856)*, 1: *Aegypten – Geten (Nr. 608a-708)* (Leiden, 1958).

Jacquart D. and Micheau, F., *La médecine arabe et l'occident médiéval* (Paris, 1990).

Janson, T., *Latin Prose Prefaces. Studies in Literary Conventions* (Göteborg/Upssala, 1964).

Jeauneau, E., 'Le commentaire érigénien sur Martianus Capella (*De nuptiis, Lib. I*) d'après le manuscrit d'Oxford (*Bodl.Libr. Auct.T.2.19, fol. 1-31*), in: *Quatres thèmes érigéniens* (Montreal/Paris, 1978), pp. 91-166.

Johnson, W.A., 'Toward a Sociology of Reading in Classical Antiquity', *American Journal of Philology* 121 (2000), pp. 593-627.

Jones, R.M., 'The Ideas as the Thoughts of God', *Classical Philology* 21 (1921), pp. 317-326.

Kaiser, R., 'Les évêques de Langres dans leur fonction de *missi dominici*', in: Semmler, J., ed., *Langres et ses évêques, VIIIe-XIe siècles: aux origines d'une seigneurie ecclésiastique, Actes du colloque Langres-Ellwangen 28 Juin 1985)* (Langres, 1986), pp. 93-111.

Kaiser, R., 'Royauté et pouvoir épiscopal au nord de la Gaule (VIIe-IXe siècles)', in: *La Neustrie. Les pays au nord de la Loire de 650 à 850*, Beihefte der Francia 16, 2 vols. (1989), vol. 1, pp. 143-160.

Keefe, S.A., 'Carolingian Baptismal Expositions: A Handlist of Tracts and Manuscripts', in: Blumenthal, U.R., ed., *Carolingian Essays: Andrew W. Mellon Lectures in Early Christian Studies* (Washington, 1983), pp. 169-237.

Keenan, J.G., 'The Names Flavius and Aurelius as Status Designations in Later Roman Egypt', *Zeitschrift für Papyrologie und Epigraphik* 13 (1974), pp. 283-304.

———, 'The Aphrodito Papyri and Village Life in Byzantine Egypt', *Bulletin de la Société d'Archéologie Copte* 26 (1984), pp. 51-63.

Kemmer, E., *Die polare Ausdrucksweise in der griechischen Literatur* (Würzburg, 1903).

Kiessling, E., ed., *Sammelbuch griechischer Urkunden aus Ägypten* 6 (Wiesbaden, 1958-1963).

Kittel, G., *et al.*, eds., *Theologisches Wörterbuch zum Neuen Testament* (Stuttgart, 1933-1978).

Kloppenborg, J.S., *The Formation of Q. Trajectories in Ancient Wisdom Traditions* (Harrisburg, 1987).

Koch-Westenholz, U., *Mesopotamian Astrology - an Introduction to Babylonian and Assyrian Celestial Divination* (Copenhagen, 1995).

Komoroczy, G., 'Berossos and the Mesopotamian Literature', *Acta Antiqua Academica Scientiarum Hungarica* 21 (1973), pp. 125-152.

Koschaker, P., *Europa und das römische Recht* (Munich, 1947).

Köster, H., *Einführung in das Neue Testament im Rahmen der Religionsgeschichte und Kulturgeschichte der hellenistischen und römischen Zeit* (Berlin/New York, 1980).

Kovelman, A.B., 'From Logos to Myth: Egyptian Petitions from the 5th–7th Centuries', *Bulletin of the American Society of Papyrologists* 28 (1991), pp. 133-152.

Kristeller, P.O, Cranz, F.L. *et al.*, eds., *Catalogus translationum et commentariorum: Mediaeval and Renaissance Latin Translations and Commentaries. Annotated Lists and Guides*, 3 vols. (Washington, 1976).

Krüger, P., *Geschichte der Quellen und Litteratur des römischen Rechts* (Munich /Leipzig, 1912).

Kuhn, C.A., *Channels of Imperishable Fire. The Beginnings of Christian Mystical Poetry and Dioscorus of Aphrodito* (New York, 1995).

Kühn, C.G., ed., *Claudii Galeni Opera Omnia*, vol. VI (Leipzig, 1823; repr. Hildesheim, 1965).

Kuhrt, A., 'Berossus' *Babyloniaka* and Seleucid Rule in Babylonia', in: Kuhrt, A. and Sherwin-White, S., eds., *Hellenism in the East: the Interaction of Greek and Non-Greek Civilizations from Syria to Central Asia after Alexander* (London, 1987), pp. 32-56.

Kümmel, H.M., *Familie, Beruf und Amt im spätbabylonischen Uruk. Prosopographische Untersuchungen zu Berufsgruppen des 6. Jahrhunderts v. Chr. in Uruk* (Berlin, 1979).

Kunkel, W., *Herkunft und soziale Stellung der römischen Juristen* (Graz/Vienna/ Cologne, ^{2}1967).

Kuttner, S., 'The revival of jurisprudence', in: Benson, R.L. and Constable, G., eds., *Renaissance and Renewal in the Twelfth Century* (Cambridge, MA, 1982), pp. 299-323.

Labourt, J., *Le christianisme dans l'empire Perse sous la dynastie Sassanide* (Paris, 1904).

Lactantius, *Divinae institutiones*, *CSEL* 19, ed. S. Brandt (Vienna, 1890).

———, *De opificio Dei*, *CSEL* 27.1, ed. S. Brandt (Vienna, 1893).

Lambdin, Th.O., transl., *The Gospel of Thomas*, in: Robinson, *The Nag Hammadi Library*, pp. 126-138.

Landsberger, B., 'The Conceptual Autonomy of the Babylonian World', *MANE* 1 (1964), pp. 64-71.

Larsen, M.T., 'The "Babel/Bible" Controversy and its Aftermath', in: Sasson, J., ed., *Civilizations of the Ancient Near East* (New York, 1995), pp. 95-106.

Le Bras, G., 'Notes pour servir à l'histoire des collections canoniques (1)', *Revue historique de droit français et étranger* 9 (1930), pp. 506-524.

Lee, A.D., *Pagans & Christians in Late Antiquity. A Sourcebook* (London, 2000).

Leeman, A.D., *Hieronymus' droom: de betekenis van Cicero voor christendom en humanisme* (Leiden, 1952).
Lemoine, F., *Martianus Capella: A Literary Reevaluation* (Munich, 1972).
Leonardi, C., 'I codici di Marziano Capella I-II', *Aevum* 33 (1959), pp. 443-489, and *Aevum* 34 (1960), pp. 1-99 and 411-524. Leonardi, C., 'Glosse eriugeniane a Marziano Capella in un codice Leidense', in: *Jean Scot Erigène et l'histoire de la philosophie* (Paris, 1977), pp. 171-182.
Lesne, É., *La hiérarchie épiscopale: provinces, métropolitains, primats en Gaule and Germanie 742-882* (Lille, 1905).
Levine, L.I., *The Rabbinic Class of Roman Palestine in Late Antiquity* (Jerusalem, 1989).
———, *The Ancient Synagogue: The First Thousand Years* (New York/London, 2000).
Lévi-Strauss, C., *Tristes tropiques* (Paris, 1955).
———, *Anthropologie structurale* (Paris, ^{2}1974).
Lewis, C. and Short, C., *A Latin Dictionary* (Oxford, 1879).
Lewis, C.S., *The Allegory of Love* (New York, 1963).
Leyser, C., *Authority and asceticism from Augustine to Gregory the Great* (Oxford, 2000).
Lieberman, S., *Hellenism in Jewish Palestine* (New York, 1962).
Liebeschütz, H., 'Zur Geschichte der Erklärung des Martianus Capella bei Eriugena', *Philologus* 104 (1960), pp. 127-137.
Liebs, D., 'Recht und rechtsliteratur', in: Herzog, R. and Schmidt, P.L., eds., *Handbuch der lateinischen Literatur der Antike*, vol. 5 (Munich, 1989), pp. 83-217.
———, 'Jurisprudenz', in: Herzog, R. and Schmidt, P.L., eds., *Handbuch der lateinischen Literatur der Antike*, vol. 4 (Munich, 1997), pp. 55-73.
Lindberg, D.C., *The Beginnings of Western Science: The European Scientific Tradition in Philosophical, Religious, and Institutional Context, 600 B.C. to A.D. 1450* (Chicago/London, 1992).
Lirinensis, Vincentius, *For the Antiquity and Universality of the Catholic Faith against the Profane Novelties of all Heretics* (Latin and English) (London, 1899).
Livrea, E., *Parafrasi del vangelio di S. Giovanni. Canto XVIII* (Naples, 1989).
Lloyd, A.C., 'The Principle that the Cause is Greater that its Effect', *Phronesis* 21 (1976), pp. 146-156.
Loenen, J.H., 'Albinus' Metaphysics. An Attempt at Rehabilitation' I, *Mnemosyne* IV/9 (1956), pp. 297-319 and II, *Mnemosyne* IV/10 [1957], pp. 35-56.
Lord, A.B., *The Singer of Tales* (Cambridge, 1960).
Lusigna, S., 'La transmission parascolaire de savoirs juridiques: les arts épistolaires de la chancellerie royale française', in: *Éducation, apprentissages, initiation au Moyen Âge* (Montpellier, 1993), vol. 1, pp. 249-262.
Luttikhuizen, G., 'The Religious Message of Andrew's Speeches', in: Bremmer, J.N., ed., *Acts of Andrew* (Louvain, 2000), pp. 96-103.
Lutz, C.E., ed., *Johannes Scottus, Annotationes in Marcianum* (Cambridge, 1939).
———, *Dunchad Glossae in Martianum*, Philological Monographs published by the American Philological Association 12 (Lancaster/Oxford, 1944).

———, ed., *Remigius Autissiodorensis, Commentum in Martianum Capellam, Libri I-II* (Leiden, 1962), *Libri III-IX* (Leiden, 1965).

———, 'Martianus Capella', in: Kristeller, Cranz *et al.*, *Catalogus translationum et commentariorum*, vol. II, pp. 367-381; see also 'Addenda et corrigenda', vol. III (1976), pp. 449-452, and vol. VI (1986), pp. 185-186.

Maassen, F., *Geschichte der Quellen und der Literatur des canonischen Rechts im Abendlande bis zum Ausgang der Mittelalters*, vol. 1 (1870; repr. Graz, 1956).

———, *Concilia aevi merovingici*, *MGH Concilia* 1 (1893; repr. Hannover, 1956).

MacCoull, L.S.B., *Dioscorus of Aphrodito. His Work and his World* (Berkeley/Los Angeles/London, 1989).

McKitterick, R., *The Frankish Kingdoms under the Carolingians, 751-987* (London, 1983; repr. 1996).

———, *The Carolingians and the Written Word* (Cambridge, 1989).

———, 'Eighth-century Foundations', in: *The New Cambridge Medieval History*, vol. 2, *c.700-c.900* (Cambridge, 1995), pp. 681-694.

Maehler, H., 'Menander Rhetor and Alexander Claudius in a Papyrus Letter', *Greek Roman and Byzantine Studies* 15 (1974), pp. 305-312.

Manitius, M., *Geschichte der lateinischen Literatur des Mittelalters*, vol. I (Munich, 1911).

———, 'Zur karolingischen Literatur', *Neues Archiv* 36 (1911), p. 43-75.

Mansfeld, J., 'Three notes on Albinus', *Theta-Pi* 1 (1972), pp. 61-80.

Marrou, H.-I., *Histoire de l'éducation dans l'antiquité* (Paris, 1948).

Maspero, J., 'Études sur les papyrus d'Aphrodité', *Bulletin de l'Institut Français d'Archéologie Orientale* 7 (1910), pp. 97-152.

———, 'Un dernier poète grec d'Egypte : Dioscore, fils d'Apollôs', *Revue des Études Grecques* 24 (1911), pp. 426-481.

———, *Papyrus grecs d'époque byzantine*, vols. 1-3 (Cairo, 1911-1916).

Mathisen, R.W., *Ecclesiastical Factionalism and Religious Controversy in Fifth-Century Gaul* (Washington, 1989).

———, 'The "Second Council of Arles" and the Spirit of Compilation and Codification in Late Roman Gaul', *Journal of Early Christian Studies*, 5 (1997), pp. 511-554.

Mayer W.R. and Van Dijk, J., *Texte aus dem Reš-Heiligtum in Uruk-Warka* (Berlin, 1980).

McCullough, J.C., 'Early Syriac Commentaries on the New Testament', *Theological Review* 5 (1982), pp. 14-33, 79-126.

McDonough, C. 'Trier, Bibliothek des Priesterseminars Ms. 100 and the Text of Martianus Capella', *Mediaeval Studies* 36 (1974), pp. 56-66.

McEwan, G.J.P., *Priest and Temple in Hellenistic Babylonia* (Wiesbaden, 1981).

Mela, Pomponius, *De situ orbis libri tres*, eds. J. and A. Gronovius (London, 1819) (available on: http://ourworld.cs.com/latintexts/mela_home_page.htm).

———, *Description of the World*, transl. F.E. Romer (Ann Arbor, 1998).

Ménard, J.É., *L'Évangile de Vérité* (Leiden, 1972).

Merlan, P., *From Platonism to Neoplatonism* (The Hague, ^{2}1960).

———, *Monopsychism, Mysticism, Metaconsciousness. Problems of the Soul in Neoaristotelianism and Neoplatonism* (The Hague, 1963).

Meyerhof, M., 'Die Augenheilkunde in der von Budge herausgegebenen syrischen ärzlichen Handschrift', *Der Islam* 6 (1916), pp. 257-268.

Michaelides, D., *Foi, écritures et tradition ou les <Praescriptiones> chez Tertullien* (Paris, 1969).

Milde, W., *Der Biblioteksкatalog des Klosters Murbach aus dem 9. Jahrhundert. Ausgabe und Untersuchung von Beziehungen zu Cassiodors 'Institutiones'* (Heidelberg, 1968).

Milham, M.E., 'C. Julius Solinus', in: Cranz, F.E., Brown, V. and Kristeller, P.O., eds., *Catalogus translationum et commentariorum: Mediaeval and Renaissance Latin Translations and Commentaries* (Washington, DC, 1986), vol. VI, pp.73-85.

Miller, T.S., *The Birth of the Hospital in the Byzantine Empire* (Baltimore/London, 1985).

Minchin, E., *Homer and the Resources of Memory. Some Applications of Cognitive Theory to the* Iliad *and the* Odyssey (Oxford, 2001).

Mitteis, L., *Grundzüge und Chrestomathie der Papyruskunde* 2.2 (Berlin/Leipzig, 1912).

Mohr, W., 'Die kirchliche Einheitspartei und die Durchführung der Reichsordnung von 817', *Zeitschrift für Kirchengeschichte* 72 (1961), pp. 1-45.

Momigliano, A., ed., *The Conflict between Paganism and Christianity in the Fourth Century* (Oxford, 1963).

———, *Alien Wisdom: The Limits of Hellenization* (Cambridge, 1975).

Mommsen, Th., Krüger, P., Schöll, R. and Kroll., W., eds., *Corpus iuris civilis, editio stereotypa*, vol. 1: *Institutiones* and *Digesta* (Berlin, 1872); vol. 2 : *Codex Iustinianus* (Berlin, 1877); vol. 3: *Novellae* (Berlin. 1895).

Moraux, P., *Alexandre d'Aphrodise. Exégète de la noétique d'Aristote* (Paris, 1942).

———, *Der Aristotelismus bei den Griechen von Andronicos bis Alexander von Aphrodisias*, II. *Der Aristotelismus im I. und II. Jh. n. Chr.* (Berlin/New York, 1984).

Mordek, H., 'Dionysio-Hadriana und Vetus Gallica – historische geordnetes und systematisches Kirchenrecht am Hofe Karls des Grossen', *Zeitschrift der Savigny-Stiftung für Rechtsgeschichte, Kanonistische Abteilung* 55 (1969), pp. 39-63.

Morgan, T.J., *Literate Education in the Hellenistic and Roman Worlds* (Cambridge, 1998).

Morin, G., 'Le *Liber dogmaticum* de Gennade de Marseille et problèmes qui s'y rattachent', *Revue bénédictine* 24 (1907), pp. 445-455.

Munier, C., 'Les sources patristiques du droit de l'Église du VIIIe au XIIIe siècle', *Revue de droit canonique* 4 (1954), pp. 184-192; repr. in: *Vie conciliaire et collections canoniques en Occident IVe-XIIe siècles* (London, 1987).

Nau, F., ed. and transl., *Barhadbešabba 'Arbaïa. Histoire ecclésiastique (IIe partie)*, *PO*, IX (Paris, 1913, repr. Turnhout, 1983).

Neugebauer, O., *A History of Ancient Mathematical Astronomy* (Berlin, 1975).

Norden, E., *Die antike Kunstprosa vom vi. Jahrhundert v. Chr. bis in die Zeit der Renaissance*, 2 vols. (Leipzig, ^{2}1909).

Nuckolls, C.W., 'Cognitive Anthropology', in: Bechtel, W. and Graham, G., eds., *A Companion to Cognitive Science* (Malden, MA/Oxford, 1998), pp. 140-145.

Nutton, V., 'From Galen to Alexander. Aspects of Medicine and Medical Practice in Late Antiquity', in: Scarborough, J., ed., *Symposium on Byzantine Medicine 1983* (Washington, 1984), pp. 1-14.

O'Leary, De Lacy, *How Greek Science Passed to the Arabs* (London/Boston/ Henley, 1980).

Oelsner, J., *Materialien zur babylonischen Gesellschaft und Kultur in hellenistischer Zeit* (Budapest, 1986).

———, 'Aus dem Leben babylonischer "Priester" in der 2. Hälfte des 1. Jahrtausends v. Chr. (am Beispiel der Funde aus Uruk)', in: Zablocka, J. and Zawadzki, S., eds., *Everyday Life in the Ancient Near East. Papers Presented at the International Conference Poznán, 19-22 September, 1989* (Poznan, 1993), pp. 235-242.

Osler, D., 'The Compilation of Justinian's Digest', *Zeitschrift der Savigny-Stiftung Rom. Abt.* 102 (1985), pp. 129-184.

P.Aphrod.Lit. = Fournet, *Hellénisme.*

P.Aphrod.Lit. = Fournet, *Hellénisme.*

P.Aphrod.Lit. = Fournet, J.-L., *Hellénisme dans l'Égypte du VI*[e] *siècle. La bibliothèque et l'œuvre de Dioscore d'Aphrodité* (Le Caire, 1999).

P.Cair.Masp. I-III = Maspero, J., *Papyrus grecs d'époque byzantine* (Le Caire 1911-1916).

P.Cair.Masp. I-III = Maspero, *Papyrus grecs.*

P.Cair.Masp. I-III = Maspero, *Papyrus grecs.*

P.Flor. I = Vitelli, *Documenti pubblici e privati.*

P.Flor. III = Vitelli, *Documenti.*

P.Hamb. III = Hagedorn and Kramer, *Griechische Papyrusurkunden.*

P.Lond. V = Bell, *Greek Papyri.*

P.Lond. V = Bell, H.I., *Greek Papyri in the British Museum*, vol. V (London, 1917).

P.Mich. XIII = Sijpesteijn, *The Aphrodite Papyri.*

P.Oxy. I = Grenfell and Hunt, *The Oxyrhynchus Papyri.*

P.Oxy. L = Parsons, Rea and Turner, *The Oxyrhynchus Papyri.*

P.Oxy. XXVII = Turner, Rea, Koenen and Pomar, *The Oxyrhynchus Papyri.*

P.Thomas = Gagos and Bagnall, *Essays and Texts.*

Papathomas, A., 'Zwischen juristischen Formeln und künstlerischer Schöpfung. Neutestamentliche Elemente in den Urkunden des spätantiken Dichters und Notars Flavius Dioskoros von Aphrodito', *Hermes* 128 (2000), pp. 481-499.

Parma, C., *Pronoia und Providentia, Der Vorsehungsbegriff Plotins und Augustins* (Leiden , 1971).

Parpola, S., *Letters from Assyrian Scholars to the Kings Esarhaddon and Assurbanipal: II Commentary and Appendices* (Neukirchen-Vluyn, 1983).

———, 'Mesopotamian Astrology and Astronomy as Domains of the Mesopotamian "Wisdom"', in: Galter, H.D., ed., *Die Rolle der Astronomie in den Kulturen Mesopotamiens. Beiträge zum 3. Grazer Morgenländischen Symposion (23.-27. September 1991)* (Graz, 1993), pp. 49-50.

———, *Letters from Assyrian and Babylonian Scholars* (Helsinki, 1993).

Parry, Milman, 'On Typical Scenes in Homer', in: Parry, A., ed., *The Making of Homeric Verse: The Collected Papers of Milman Parry* (New York/Oxford, 1987), pp. 404-407.

Parsons, P.J., Rea, J.R. and Turner, E.G., eds., *The Oxyrhynchus Papyri*, vol. L (London, 1983).

Partsch, J., 'Neue Urkunden zum justinianischen Reskriptenprozesse', *Nachrichten von der K. Gesellschaft der Wissenschaften zu Göttingen* (1911), pp. 201-253.

Payne Smith, R., *Thesaurus Syriacus* (Oxford, 1879-1901).

Pearce, L. and Doty, T., 'The Activities of Anu-belšunu, Seleucid Scribe', in: Marzahn, J. and Neumann, H., eds., *Assyriologica et Semitica. Festschrift für Joachim Oelsner anläßlich seines 65. Geburtstages am 18. Februar 1997* (Münster, 2000), pp. 331-341.

Pedersén, O., *Archives and Libraries in the Ancient Near East 1500-300* (Bethesda, 1998).

Pelikan, J., *The Christian Tradition: A History of the Development of Doctrine*, vol. 1: *The Emergence of the Catholic Tradition (100-600)* (Chicago, 1971).

Pingree, D., 'Legacies in Astronomy and Celestial Omens', in: Dalley, S. ed., *The Legacy of Mesopotamia* (Oxford, 1998), pp. 125-137.

Pliny, *Naturalis historiae libri xxxvii*, eds. and transl. H. Rackham (libri i-xix, xxxiii-xxxv), W.H.S. Jones (libri xx-xxxii) and D.E. Eichholz (libri xxxvi-xxxvii), 10 vols. (Cambridge, 1949-1963).

PO: *Patrologia Orientalis*.

Pongratz-Leisten, B., *Herrschaftswissen in Mesopotamien* (Helsinki, 1999).

Préaux, J., 'Le commentaire de Martin de Laon sur l'œuvre de Martianus Capella', *Latomus* 12 (1953), pp. 437-459.

———, 'Un nouveau manuscrit de Saint-Gall: le Bruxellensis 9565-9566', *Scriptorium* 10 (1956), pp. 221-228.

———, 'Deux manuscrits gantois de Martianus Capella', *Scriptorium* 13 (1959), pp. 15-21.

———, 'Les manuscrits principaux du *De nuptiis Philologiae et Mercurii* de Martianus Capella', in: *Lettres latines du moyen âge et de la renaissance*, Collection Latomus 158 (Brussels, 1987), pp.76-128.

Preisendanz, K., *Papyri graecae magicae* 2 (Stuttgart, [2]1974).

Pricoco, S., *Storia letteraria e storia ecclesiastica dal De viris illustribus di Girolamo a Gennadio* (Catania, 1979).

PSI VIII = *Papiri greci e latini* 8 (Florence, 1927)

Puech, H.-C. and Blatz, B., 'Other Gnostic Gospels and Related Literature', in: Schneemelcher, W. and McL. Wilson, R., eds., *New Testament Apocrypha*, vol. 1 (Cambridge/Louisville, [2]1991), pp. 354-413.

Pugsley, D., 'The Florentine Index', *Journal of Legal History* 14 (1993), pp. 94-105 (= Pugsley, D., *Justinian's Digest and the Compilers* (Exeter, 1995), pp. 41-52).

Putman, H., *L'église et l'islam sous Timothée I (780-823). Étude sur l'église nestorienne au temps des premiers 'Abbāsides avec nouvelle édition et traduction du dialogue entre Timothée et al-Mahdi* (Beirut, 1975).

Quasten, J., *Patrology*, 4 vols. (Westminster, 1986).

Rabello, A.M., *The Jews in the Roman Empire: Legal Problems from Herod to Justinian* (Aldershot, 2000).

Reale, G., *A History of Ancient Philosophy*, vol. IV. *The Schools of the Imperial Age* (New York, 1990).

Reif, S., *Judaism and Hebrew Prayer* (Cambridge, 1993).

Reinink, G.J., 'Seharbokt, ein nestorianischer Lehrer der islamischen Zeit', in: *Symposium Syriacum 1976* (Rome, 1978), pp. 73-98.

———, *Studien zur Quellen- und Traditionsgeschichte des Evangelien-kommentars der Gannat Bussame, CSCO,* 414, *Subsidia*, 57 (Louvain, 1979).

———, 'Der Dämon "Sohn des Daches" in der syrischen exegetischen Literatur', in: Livingstone, E.A., ed., *Studia Patristica*, Vol. XVI, Part II (Berlin, 1985), pp. 105-113.

———, ed. and transl., *Gannat Bussame. I. Die Adventssonntage, CSCO*, 510, 502, *Scriptores Syri*, 211, 212 (Louvain, 1988).

———, 'De invloed van de antieke geneeskunde op de bijbelexegese van Seharbokt bar Mesargis', in: Van Dijk, H. and Smits, E.R., eds., *Dwergen op de schouders van Reuzen. Studies over de receptie van de Oudheid in de Middeleeuwen* (Groningen, 1990), pp. 31-42.

———, '"Edessa grew dim and Nisibis shone forth": the School of Nisibis at the transition of the sixth-seventh century', in: Drijvers, J.W. and MacDonald A.A., eds., *Centres of Learning: Learning and Location in Pre-Modern Europe and the Near East* (Leiden/New York/Cologne, 1995), pp. 77-89.

Relihan, J.C., *Ancient Menippean Satire* (Baltimore, 1993).

Reynolds, R.E., 'The Visigothic Liturgy in the Realm of Charlemagne', in: Berndt, R., ed., *Das Frankfurter Konzil von 794. Kristallisationspunkt karolingischer Kultur*, 2 vols. (Mainz, 1997), vol. 2, pp. 919-945.

Reznikoff, I., 'Le chant des Gaules sous les carolingiens' in: Sot. M., ed., *Haut moyen-âge. Culture, éducation et société. Études offertes à Pierre Riché* (La Garenne-Colombes, 1990), pp. 323-342.

Riccobono, S., ed., *Fontes Juris Romani Antejustiniani I* (Florence, [2]1941).

Rich, A.N.M., 'The Platonic Ideas as the Thoughts of God', *Mnemosyne* IV 7 (1954) 123-133.

Richardson, E.C., ed., *Gennadius, De viris inlustribus* (Leipzig, 1896).

Riché, P., 'Les centres de culture en Neustrie de 650 à 850', in: *La Neustrie. Les pays au nord de la Loire de 650 à 850*, Beihefte der Francia 16, 2 vols. (1989), vol. 2, pp. 297-305.

Rives, R., 'Human Sacrifice Among Pagans and Christians', *Journal of Roman Studies* 85 (1995), pp. 65-85.

Robinson, J.M., ed., *The Nag Hammadi Library in English* (San Francisco, [4]1996).

Rochberg, F., 'The Cultural Locus of Astronomy in Late Babylonia', in: Galter, H.D., ed., *Die Rolle der Astronomie in den Kulturen Mesopotamiens. Beiträge zum 3. Grazer Morgenländischen Symposion (23.-27. September 1991)* (Graz, 1993), pp. 31-45.

———, *Babylonian Horoscopes* (Philadelphia, 1998).

———, 'Babylonian Horoscopy: The Texts and their Relations', in: Swerd-low, N.M. ed., *Ancient Astronomy and Celestial Divination* (Cambridge, MA, 1999), pp. 39-59.

———, 'Scribes and Scholars: The *ṭupšar Enuma Anu Enlil*', in: Marzahn, J. and Neumann, H., eds., *Assyriologica et Semitica. Festschrift für Joachim Oelsner anläßlich seines 65. Geburtstages am 18. Februar 1997* (Münster, 2000), pp. 359-375.

Rochberg-Halton, F., 'Between Observation and Theory in Babylonian Astronomical Texts', *Journal of Near Eastern Studies* 50 (1991), pp. 107-120.

Rutgers, L.V., 'Attitudes to Judaism in the Greco-Roman Period: Reflections on Feldman's *Jew and Gentile in the Ancient World*', in: Rutgers, *The Hidden Heritage*, pp. 199-234.

———, *The Hidden Heritage of Diaspora Judaism* (Louvain, 1998).

Sachs, A., *Late Babylonian Astronomical and Related Texts, copied by T.G. Pinches and J.N. Strassmaier* (Providence, 1955).

Sachs, A. and Hunger, H., *Astronomical Diaries and related texts from Babylonia.* vol. 1: *Diaries from 652 B.C. to 262 B.C*; vol 2: *Diaries from 261 B.C. to 165 B.C.*; vol. 3: *Diaries from 164 B.C. to 61 B.C.* (Vienna, 1988, 1989, 1996).

SB VI = Kiessling, *Sammelbuch griechischer Urkunden.*

SB XIV = *Sammelbuch griechischer Urkunden aus Aegypten* (Wiesbaden, 1981-1983).

Schachter, D.L., ed., *Memory Distortion: How Minds, Brains, and Societies Reconstruct the Past* (Cambridge/London, 1995).

Schäfer, P., *The Talmud Yerushalmi and Graeco-Roman Culture* II, Texts and Studies in Ancient Judaism 79 (Tübingen, 1998).

Schank, R.C. and Abelson, R.P., *Scripts, Plans, Goals and Understanding: An Inquiry into Human Knowledge Structures* (Hillsdale, 1977).

Scheindler, A., *Nonni Panopolitani Paraphrasis S. Evangelii Ioannei* (Leipzig, 1881).

Scheltema, H.J., 'Das Kommentarverbot Justinians', *Tijdschrift voor Rechtsgeschiedenis* 45 (1977), pp. 307-331.

Scheltema, H.J., Holwerda, D. and Van der Wal, N., eds., *Basilicorum libri LX*, 17 vols. (Groningen, 1953-1988).

Schenke, H.M., 'The Gospel of Philip', in: Schneemelcher and Wilson, *New Testament Apocrypha*, vol. 1, pp. 179-208.

Scher, A., ed. and transl., *Mar Barhadbšabba 'Arbaya.Cause de la fondation des écoles*, *PO*, IV (Paris 1907; repr. Turnhout, 1971).

———, *Histoire nestorienne inédite (Chronique de Séert)*, *PO*, XIII (Paris, 1918; repr. Turnhout, 1973).

Schieffer, R., 'Spätantikes Kirchenrecht in einer rätischen Sammlung des 8. Jahrhunderts', *Zeitschrift der Savigny-Stiftung für Rechtsgeschichte, Kanonistische Abteilung* 66 (1980), pp. 164-191.

Schmidt, P.L., 'Solins Polyhistor in Wissenschaftsgeschichte und Geschichte', *Philologus* 139 (1995), pp. 23-35.

Schneemelcher, W., 'Gospels. Non-Biblical Material about Jesus', in: Schneemelcher, W. and Wilson, R. McL., eds., *New Testament Apocrypha*, 2 vols. (Cambridge/Louisville, [2]1991-1992), vol. 1, pp. 77-87.

Schöffler, H.H., *Die Akademie von Gondischapur: Aristoteles auf dem Wege in den Orient* (Stuttgart, [2]1980).

Schrimpf, G., 'Zur Frage der Authentizität unserer Texte von Johannes Scottus' *Annotationes in Martianum*', in: O'Meara, J.J. and Bieler, L., eds., *The Mind of Eriugena* (Dublin 1973), pp. 125-139.

Schroeder, F.M., 'The Analogy of the Active Intellect to Light in *De Anima* of Alexander of Aphrodisias', *Hermes* 109 (1981), pp. 215-225.

Schulz, F., *History of Roman Legal Science* (Oxford, ²1953).
Schwartz, E., ed., *Acta Conciliorum Oecumenicorum* (Berlin, 1914-1983).
———, 'Die Kanonessammlungen der alten Reichskirche', *Zeitschrift der Savigny-Stiftung für Rechtsgeschichte, Kanonistische Abteilung* 25 (1936), pp. 1-114.
Schwartz, S., 'Language, Power and Identity in Ancient Palestine', *Past and Present* 108 (1995), pp. 3-47.
Schweigert, W., 'Die Theologenschule von Bēt Lāpāt – Ğundaisābūr. Ein Beitrag zur nestorianischen Schulgeschichte', *Zeitschrift der Deutschen Morgenländischen Gesellschaft*, Suppl. 4: *XX. Deutscher Orientalistentag 1977 in Erlangen* (Wiesbaden, 1980), pp. 185-187.
———, *Das Christentum in Hūzistān im Rahmen der frühen Kirchengeschichte Persiens bis zur Synode von Seleukia-Ktesiphon im Jahre 410* (Marburg, 1989).
Seneca, *Ad Lucilium epistulae morales* (Latin and English), R.M. Gummere, 3 vols. (Cambridge, 1925; repr. 1971).
Sezgin, F., *Geschichte des arabischen Schrifttums*, III (Leiden, 1970).
Shanzer, D., 'Review Article: Felix Capella: minus sensus quam nominis pecudalis', *Classical Philology* 81 (1986), pp. 62-81.
Sharpless, R.W., 'Alexander of Aphrodisias: Scholasticism and Innovation', *Aufstieg und Niedergang der römischen Welt* II.36.2 (New York/Berlin, 1987), pp. 1178-1243.
Shils, E., *Tradition* (Chicago, 1981).
Shorrock, R., *The Challenge of Epic. Allusive Engagement in the Dionysiaca of Nonnus* (Leiden, 2001).
Sijpesteijn, P.J., *The Aphrodite Papyri in the University of Michigan Papyrus Collection* (Zutphen, 1977).
Solinus, C. Iulius, *Collectanea rerum memorabilium*, ed. Th. Mommsen (Berlin, ²1895).
Sperber, D., *Magic and Folklore in Rabbinic Literature* (Ramat-Gan, 1994).
Steele, J., 'A Simple Function for the Length of the Saros in Babylonian Astronomy', in: Steele, J.M. and Imhansen, A., eds., *Under One Sky – Astronomy & Mathematics in the Ancient Near East*. Forthcoming.
Stein, P.G., 'Legal Humanism and Legal Science', *Tijdschrift voor Rechtsgeschiedenis* 54 (1986), pp. 297-306 (= Stein, P.G., *The Character and Influence of the Roman Civil Law* (London/Ronceverte, 1988), pp. 91-100).
Stemberger, G., *Introduction to the Talmud and Midrash* (Edinburgh, 1996).
———, *Jews and Christians in the Holy Land, Palestine in the Fourth Century* (Edinburgh, 2000).
Stern, M., *Greek and Latin Authors on Jews and Judaism*, 3 vols. (Jerusalem, 1974-1984).
Stirnimann, J.K., *Die Praescriptio Tertullians im Lichte des Römischen Rechts und der Theologie* (Freiburg, 1949).
Stol, M., *Epilepsy in Babylonia* (Groningen, 1993).
Stolte, B.H., 'Justinianus Bifrons', in: Magdalino, P., ed., *New Constantines. The Rhythm of Imperial Renewal in Byzantium, 4th-13th Centuries* (Aldershot, 1994), pp. 45-55.
Strewe, A., ed., *Die Canonessammlung des Dionysius Exiguus in der ersten Redaktion* (Berlin, 1931).

Sullivan, R.E., 'The Context of Cultural Activity in the Carolingian Age', in: Sullivan, R.E., ed., *The Gentle Voices of Teachers. Aspects of Learning in the Carolingian Age* (Ohio, 1995), pp. 51-105.

Talon, P., '*Enūma eliš* and the Transmission of Babylonian Cosmology to the West', in: Whiting, R.M., ed., *Mythology and Mythologies: Methodolo-gical Approaches to Intercultural Influences - Proceedings of the Second Annual Symposium of the Assyrian and Babylonian Intellectual Heritage Project Held in Paris, France, October 4-7, 1999* (Helsinki, 2001).

Teeuwen, M., 'Harmonie der hemelsferen. Een klassiek thema geïnterpreteerd door Johannes Scotus Eriugena', *Millennium* 9 (1995), pp. 117-132.

———, *Harmony and the Music of the Spheres. The Ars Musica in Ninth-Century Commentaries on Martianus Capella* (Leiden/Boston/Cologne, 2002).

Tertullian, *The Writings of Quintus Sept. Flor. Tertullianus*, 3 vols. in: Roberts, A. and Donaldson, J., eds., *Ante-Nicene Christian Library: Translations of the Writings of the Fathers Down to A.D. 325*. vols. XI, XV, XVIII (Edinburgh, 1895).

———, *Le premier livre Ad nationes de Tertullien. Introduction, texte, tra-duction et commentaire*, ed. A. Schneider (Rome, 1968).

The Holy Bible Containing the Old and New Testaments. New Revised Standard Version (Grand Rapids, MI, 1989).

Theiler, W., 'Die Entdeckung der Metaphysik des Aristoteles', *Museum Helveticum* 15/2 (1958) 85-105.

Thieme, H., *Das Naturrecht und die europäische Rechtsgeschichte* (Basel, [2]1954).

Thomassen, E., *Le Traité Tripartite (NHI I, 5)* (Québec, 1989).

Thureau-Dangin, F., *Tablettes d'Uruk à l'usage des prêtres du Temple d'Anu au temps des Séleucides* (Paris, 1922).

Tisserant, E., 'L'église nestorienne', in: *Dictionnaire de théologie catholique*, vol. 11, cols. 157-323.

Trapp, M.B., *Maximus Tyrius. Dissertationes* (Stuttgart/Leipzig, 1994).

———, *Maximus of Tyre. The Philosophical Orations* (Oxford, 1997).

Traube, L., 'Computus Helperici', *Neues Archiv* 18 (1893), pp. 103-104.

Troje, H.E., *Graeca leguntur. Die Aneignung des byzantinischen Rechts und die Entstehung eines humanistischen Corpus iuris civilis in der Jurisprudenz des 16. Jahrhunderts* (Cologne, 1971).

Tulving, E., 'Episodic and Semantic Memory', in: Tulving, E. and Donaldson, W., eds., *Organization of Memory* (New York/London, 1972), pp. 381-403.

———, 'Memory: Performance, Knowledge, and Experience', *The European Journal of Cognitive Psychology* 1 (1989), pp. 3-26.

Tulving, E. and Craik, F.I.M., eds., *The Oxford Handbook of Memory* (Oxford, 2000).

Turner, E.G., Rea, J., Koenen, L. and Pomar, J.M.F., eds., *The Oxyrhynchus Papyri*, Vol. XXVII (London, 1962).

Turner, M.L., *The Gospel According to Philip: The Sources and Coherence of an Early Christian Collection* (Leiden/New York, 1996).

Ullmann, M., *Die Medizin im Islam* (Leiden /Köln, 1970).

Urbach, E.E., 'Class-Status and Leadership in the World of the Palestinian Sages', *Proceedings of the Israel Academy of Sciences and Humanities* II, No. 4 (Jerusalem, 1966), pp. 1-37.

Uro, R., 'Thomas and Oral Gospel tradition', in: Uro, R., ed., *Thomas at the Crossroads: Essays on the Gospel of Thomas* (Edinburgh, 1998), pp. 8-32.

Valantasis, R., *The Gospel of Thomas* (London/New York, 1997).

Van der Spek, R.J., 'The Babylonian Temple during the Macedonian and Parthian Domination', *Bibliotheca Orientalis* 42 (1985), pp. 541-562.

Van der Wal, N. and Lokin, J.H.A., *Historiae iuris graeco-romani delineatio. Les sources du droit byzantin de 300 à 1453* (Groningen, 1985).

Van Dijk, J., 'Die Inschriftenfunde, II. Die Tontafeln aus dem resh-Heiligtum', in: Lenzen, H.J., ed., *XVIII. vorläufiger Bericht über die von den Deutschen Archäologischen Institut und der deutschen Orient-Gesellschaft aus Mitteln der deutschen Forschungsgemeinschaft unternommenen Ausgrabungen in Uruk-Warka (1959/1960)* (Berlin, 1962), pp. 43-61.

Van Minnen, P., 'Isocrates and Menander in Late Antique Perspective', *Greek, Roman and Byzantine Studies* 33 (1992), pp. 87-98.

Van Rompay, L., 'La littérature exégétique syriaque et le rapprochement des traditions syrienne-occidentale et syrienne-orientale', *Parole de l'Orient* 20 (1995), pp.221-235.

———, 'Development of Biblical Interpretation in the Syrian Churches of the Middle Ages', in: Sæbø, M., ed., *Hebrew Bible/Old Testament: The His-tory of Its Interpretation*, Part 2 (Göttingen, 2000), pp. 559-577.

Veldhuis, N.C., *Elementary Education at Nippur – the Lists of Trees and Wooden Objects* (Groningen, 1997).

Verbeeck, B., 'Dioskoros: dorpshoofd, dichter en notaris', in: Pestman, P.W., ed., *Familiearchieven in het land van Pharao* (Zutphen, 1989), pp. 139-162.

Verbrugghe G.P., and Wickersham, J.M., *Berossos and Manetho, Introduced and Translated. Native Traditions in Ancient Mesopotamia and Egypt* (Ann Arbor, 1996).

Vessey, M., 'The Forging of Orthodoxy in Latin Christian Literature: A Case Study', *Journal of Early Christian Studies* 4 (1996), pp. 495-513.

Vian, F., *Nonnos de Panopolis. Les Dionysiaques* (Paris, 1976).

Viljamaa, T., *Studies in Greek Encomiastic Poetry of the Early Byzantine Period* (Helsinki, 1968).

Vitelli, G., *Documenti pubblici e privati dell'età romana e bizantina* (Milan, 1906).

———, *Documenti e testi letterarii dell'età romana e bizantina* (Milan, 1915).

Voetius, G., *Oratio inauguralis de pietate cum scientia conjungenda* (Utrecht, 1634).

Vogel, C., *La réforme cultuelle sous Pépin le Bref et sous Charlemagne (deuxième moitié du VIIIe siècle et premier quart du IXe siècle)*, bound with E. Patzelt, *Die karolingische Renaissance* (Graz, 1965), pp. 171-242.

Von Albrecht, M., *A History of Roman literature: From Livius Andronicus to Boethius With Special Regard to its Influence on World Literature*, 2 vols. (Leiden, 1997).

Von Dobschütz, E., ed., *Das Decretum Gelasianum De Libris recipiendis et non recipiendis in kritischen Text* (Leipzig, 1912).

Von Dzialowski, G., *Isidor und Ildefons als Literarhistoriker* (Münster, 1898).

Von Martels, Z.R.W.M., 'Between Pliny the Elder and *Altertumswissenschaft*: the style, contents and meaning of Solinus'. Forthcoming.

———, 'Solinus and Augustine: Marvels in the Book of Nature'. Forthcoming.

Von Weiher, E., *Spätbabylonische Texte aus Uruk*, vol.1 (Berlin, 1983).
———, *Spätbabylonische Texte aus Uruk*, vol. 2 (Berlin, 1988).
———, *Uruk. Spätbabylonische Texte aus dem Planquadrat U 18*, vol. 4 (Mainz am Rhein, 1993); vol. 5 (Mainz am Rhein, 1998).
Vööbus, A., ed. and transl., *The Statutes of the School of Nisibis* (Stockholm, 1961).
———, *History of the School of Nisibis*, *CSCO*, 266, *Subsidia*, 26 (Louvain, 1965).
Vorster, W.S., 'Gospel Genre', in: Freedman, D.N., *et al.*, eds., *The Anchor Bible Dictionary* (New York, 1992) vol. 2, pp. 1077-1079.
Wagener, K.C., 'Repentant Eve, Perfected Adam. Conversion in the *Acts of Andrew*', *SBL Seminar Papers* 1991 (Atlanta, 1997), pp. 348-356.
Wallace-Hadrill, J.M., *The Frankish Church* (Oxford, 1983).
Wallenfels, R., *Seleucid Archival Texts in the Harvard Semitic Museum* (Groningen, 1998).
Ward, W., *The Life of John Henry Cardinal Newman Based on his Private Journals and Correspondence*, 2 vols. (New York/London, 1912).
Wei, L., ed., *The Bilingualism Reader* (London/New York, 2000).
Werner, K.F., *Naissance de la noblesse* (Paris, 1998).
West, M.L., *The east face of Helikon: West Asiatic elements in early poetry and myth* (Oxford, 1997).
Wilmart, A., 'Les manuscrits des "Confessions" de S. Augustin. Répertoire méthodique', *Revue bénédictine* 41 (1929), pp. 325-341.
———, 'Reg. Lat. 69 (fol. 116-122): Un florilège carolingien sur le symbolisme de cérémonies du baptême, avec un appendice sur la lettre de Jean Diacre', in: Wilmart, A., *Analecta Reginensia* (Città del Vaticano, 1933), pp. 153-179.
———, 'La tradition des grands ouvrages de Saint Augustin', in: *Miscellanea Agostiniana*, 2 vols. (Rome, 1931), pp. 257-315.
Wilson, R.R., *Genealogy and History in the Biblical World* (New Haven, 1977).
Witt, R., *'In the Footsteps of the Ancients'. The Origins of Humanism from Lovato to Bruni* (Leiden, 2000).
Witt, R.E., *Albinus and the History of Middle Platonism* (Cambridge, 1937).
Wolff, H.J., 'Der byzantinische Urkundenstil Ägyptens im Lichte der Funde von Nessana und Dura', *Revue Internationale des Droits de l'Antiquité* III 8 (1961), pp. 115-154.
Wolfson, H.A., *Philo: Foundations of Religious Philosophy in Judaism, Christianity, and Islam* (Cambridge, 1948).
Wurm. H., *Studien und Texte zur Dekretalensammlung des Dionysius Exiguus* (Bonn, 1939).
Zeller, E., *Die Philosophie der Griechen in ihrer geschichtlichen Entwicklung*, $III/1^3$ (Leipzig, 1880).
Zettler, R.L., 'Written Documents as Excavated Artifacts and the Holistic Interpretation of the Mesopotamian Archaeological Record', in: Cooper, J.C. and. Schwartz, G.M., eds., *The Study of the Ancient Near East in the Twenty-First Century. The William Foxwell Allbright Centennial Conference* (Winona Lake, 1996), pp. 81-101.
Zilliacus, H., *Zur Abundanz der spätgriechischen Gebrauchssprache* (Helsinki, 1967).

INDEX

Aba, Nestorian catholicos, 164
Isaias, Abbot, 155
Abraham of Bet Rabban, 165-166
Abu Nuwas, poet, 168
Abydenus, historian, 15
Accursius, 93; *glossa ordinaria*, 93, 97
Achilles, 105-106
Acts of Andrew, 38, 44
Acts of Pilate; see *Gospel of Nicodemus*
Adam, 45-47
Adarshabur, 173
Aelianus, 69; *De natura animalium*, 69
Aeschines, 151, 153
Aesop, 151, 153
Agricola, Rudolf, 79
Alcinous, 40-45; *Didasc.*, 41
Alexander of Aphrodisias, 37-38, 40-42
Alexander the Great, 13
Alexandria, 154-155; school of, 137
al-Ma'mun, caliph, 170-171
al-Mansur, caliph, 170
Amalarius of Trier, 183
Ambrose, 179, 183
Ammianus Marcellinus, 66
Anastasia; see Tekrompia
Andrew, apostle, 47-48
Antaeopolis, 130
Anthologia Palatina, 137
Anthropos, 48
Antinoopolis, capital of Upper, 114-115, 117, 129, 130, 140, 146, 149-150, 152-157, 160
Antiochus I, Seleucid king, 13
Antu, god, 19, 22
Anu, god, 19-20, 22
Anu-aba-uṭer, son of Anu-bel-šunu, 19
Anu-aha-ušabši, son of Kidin-Anu, 19
Anu-bel-šunu, 19, 23
Aphrodite/o (Aphrodito), city, 101, 110, 115-117, 119, 123, 127-128, 130, 147-150, 152, 154, 157, 160-161
Aphroditopolis, 150
Apions, family, 154
Apollinarius of Laodicea, 105
Apollo, god, 107-108
Apollonius Rhodius, 151, 153, 160
Apollopolis, 150
Apollos, father of Dioscorus, 120-125, 128-129, 137
Arcadius Charisius, jurist, 84
Aristides, 151
Aristophanes, 151-153, 159-160
Aristotle, 10, 37-45, 48-50; *De anima*, 38-43, 46-50, 63; *Metaphysics*, 39, 41-42, 49; *De generatione animalium*, 39, 63; *Nichomachaean Ethics*, 49; *De coelo*, 41; *Eudemian Ethics*, 43, 49; *Physics*, 49; *Posterior Analytics*, 50; *Republic*, 39
Arius Didymus, 45
Arnobius, 64; *Adversus nationes*, 64
Arsinoe, 152
Artaxerxes, 77
ašipu, exorcists, 16-17, 21-23
Aspasius, 45
Asturius, bishop of Toledo, 178
Athanasius, Theban duke, 109
Atrahasis, text, 17
Augustine, 65-66, 73, 79, 175-177, 179, 183-184; *De utilitate credendi*, 65; *De consensu evangelistarum*, 73; *Confessiones*, 73, 176, 184
Aurasius, bishop, 178

Babai the Great, 167

Babylonian temple, 13
Barhadbeshabba, 164-166; *Cause of the Foundation of the Schools*, 164; *Ecclesiastical History*, 165
bāru, exispicers, 10
Basilica, 92
Basilius of Caesarea, 155
Basilius Seleuceius, 155
Bel, god of the Esagila, 23
Bellerophon, 106
Berossos/sus, 5, 13-15, 23; *Babyloniaca*, 5, 13, 15
bēt mardūtā, school, 164, 170
Bethel, city, 145
Bible translation of Aquila, 60
Birway, 166
Bit-Resh, city, 15-16, 19
Blemmyes, 109, 110, 141, 144-145
Boethius, 79, 191-192; *De institutione musica*, 192; *De consolatione philosophiae*, 79
Boktisho' d-Miles, physician, 173
Boniface, bishop, 182
Bororo Indians, 36
Browne, Sir Thomas, 63; *Pseudodoxia*, 63

Caesar, Julius, 67, 73, 75; *De bello civili*, 73
Caesarea, city, 58
Callimachus, 151, 153, 159-160
Cannibal, 'Eater of Raw Meat', 141-142, 144
Canon Muratori, 26
Cassiodorus, 181; *Institutes*, 181
Cephas, 31, 33
Chalcidius, 191
Chaldean kings, 5
Chariton, 148, 151
Charlemagne, 180-181, 183
Charles the Bald, king of France, 186, 191-193
chreiai, 33
Christodorus of Coptos, 103; *Ekphrasis* (*Palatine Anthology II*), 103
Chronicle of Seert, 167
Chrysostomus, see John
Cicero, 64, 71-74, 85, 150-151, 153; *De legibus*, 74; *De officiis*, 72; *De Oratore*, 74; *De natura deorum/The Gods*, 64, 71-72
Claudian, poet, 114
Codex des Visions, 105
Colluthus of Lycopolis, 103; *The Rape of Helen*, 103
Colossians, 156
Constantine I, emperor, 58
Constantius II, emperor, 66
Corippus, poet, 91
Council of Aachen, 183; Arles, 179; Bordeaux, 179; Chalcedon, 136, 179; Frankfurt, 175, 182-183; Lyon, 180; Mainz, 181; Milan; 94; Orange, 179; Orléans, 180; Paris, 184
Cranmer, archbishop, 65
Cutha, city, 21
Cyprian, 177
Cypris, 108
Cyril of Alexandria, 155
Cyrus of Panapolis, 103, 114; *The Palatine Anthology*, 103, 151
Cyrus, poet, 147, 161

Dan, city, 145
Daphne, 107-108
David, king of Israel, 33
Decretal of pseudo-Gelasius, 176, 181
Deir-el-Gizaz, 149
Demosthenes, 151, 153, 158
Dendera, 152-153, 157
Didymus the Blind, 155
Digesta/Digest, 83-84, 86, 88-98
Dionysius Exiguus, 176, 179, 181; *Dionysiana*, 179; *Dionysio-Hadriana*, 181
Dionysius, god, 106, 143
Dioscorus, 101-112, 115-130, 135-142, 144-148, 154, 160-161; *ethopoiiai*, 117; his mother, 124-125
Dioscorus, cousin of Dioscorus, 120-121
Dioscurides, 151

Diospolis, 148-150, 152-153, 155, 157, 160
Donne, John, 71; 'On Religion', 71
Dunchad, bishop, 187-188, 191, 192
Dura-Europos, city, 58
Dur-Sharrukin, city (Khorsbad in Irak), 16

Eanna, temple, 16, 19
Ebabbar, temple, 17, 22
Edom, kingdom, 57
Ekurzakir, 19, 21, 23
Elementa, see *Institutiones* 83
Elipandus of Toledo, 182
Emeslam, temple, 21
Enuma Anu Enlil, list of omens, 17, 19-20
Enuma Elish, epic, 16-18
Esagila, temple, 21, 23
Essenes, 77-78
Eudemus, 50
Eugenius, bishop of Toledo, 178, 183
Eunapus, 102; *Life of Sophists*, 103
Euripides, 113, 151-154, 158-159, 160; *Phoenissae*, 113
Eusebius of Caesarea, 77, 155, 177; *Chronica*, 77
Eutropius, 76
Eve, 45- 47
Ezekiel, prophet, 33

Falcidian portion, 129
Fayum, region, 149-150, 152-157
Felix of Urgel, 182
Firmicus Maternus, 78; *De errore profanarum religionum*, 78
First Epistle of Clement, 175
Flavius Aion II, 154
Frederick II, king of Prussia, 90
Fulgentius, saint, 183, 191

Gabriel bar Boktisho, director of the hospital in Jundishapur, 168
Gabriel, Christian, 166
Gabriel, student, 164,
Gabrielius, 110
Gaius, author, 148
Gaius, jurist, 82, 86-87
Galatians, 156
Galen, 151, 163-164, 169, 173; *On the Powers of Food*, 173
Galilee, 55
Galilee, Lake, 55
Gelasius I, pope, 176; *De libris recipiendis et non recipiendis*, 176
Gellius, Aulus, 69, 74; *Noctes Atticae*, 69, 74
Genizah, 148, 155, 160
Gennadius of Marseille, 177
George bar Gabriel bar Boktisho, physician, 168, 170
Giddal, rabbi, 59
Gilgamesh, epic, 16-17, 23
Goal-Year texts, 11
Golding, Arthur, 64, 66-67, 70, 75-78; *The Excellent and Pleasant Worke of Collectanea rerum memorabilium of Gaius Julius Solinus Polyhistor ...*, 64, 67, 70, 75-78
Gomorrah, 77
Gospel of Mark, 25
Gospel of Nicodemus, 26 (= *Acts of Pilate* / 'Acts of Pontius Pilate')
Gospel of Peter, 26
Gospel of Philip, 26, 35
Gospel of Thomas, 26-27, 33, 36
Gospel of Truth, 25-26
Gregory of Nyssa, 155
Grotius, Hugo, 96

Haimo of Auxerre, 188
Hammurapi, Babylonian king, 1, 17; Law Code, 1
Harun ar-Rashid, caliph, 168
Heiric of Auxerre, 188, 190
Heliodorus, 151
Henana, school of, 166, 169
Herakleopolite, 150, 152, 154-155, 157
Herennius Modestinus, jurist, 84
Hermogenianus, jurist, 84
Hermoupolis Magna, 149-150, 152-157

Herodotus, 63
Hesiod, 151, 153
Hieronymus, 64, 77
Hilary, saint, 183
Hildegard of Bingen, 73
Hillel, 54
Hippocrates, 151, 163-164
Hippolytus of Rome, 155
Homer, 27-29, 34, 101, 113, 136, 148, 151-153, 158; *Iliad*, 27-28, 113, 136-137
Hrabanus Maurus, 73

Ildefonse of Toledo, 178, 181, 183; *On Illustrious Men*, 178
Ina-qibit-Ani, 19
Index Florentinus, 85
Institutiones, 83, 88, 93
Iqiša, son of Ištar-šuma-ereš, 21-22
Irenaeus, 26; *Against Heresies*, 26
Irigal, temple, 16
Isa b. Masarjis, brother of Seharbokt (?), 168
Isa b. Sahrabuht, son of Seharbokt, 168
Isaac, 108
Isaias, prophet, 155
Isho' bar Nun, Nestorian exegetes, 169
Isho'dad of Merv, Nestorian exegetes, 169
Isho'denah of Basra, 167; *Book of Chastity*, 167
Isidore of Seville, 67, 177-178, 183, 191
Isocrates, 136-137, 151, 153
Ištar Gate, 1
Ištar-šuma-ereš, 22
Itti-Marduk-balaṭu, son of Iddin-Bel, 23
Izla, mount, 167

Jacob, 33
James, apostle, 26, 31, 33
Jericho, 77
Jeroboam, king of Israel, 144-145
Jerome, saint, 175, 177, 179, 183; *On Illustrious Men/De viris illustribus*, 177-178, 181
John Chrysostom, 77, 155
John the Scot, 186, 188-190, 192
John, apostle, 25, 28, 155, 168, 173
Joseph, Nestorian catholicos, 166
Josephus, 151
Judas Maccabaeus, 143
Julian, 183
Julian, pagarch, 122
Julianus Apostata, emperor, 76
Jundishapur, city, 163-170, 172, 173
Jurjis, 168
Justin Martyr, 26
Justinian I, emperor, 58, 82-85, 88-98, 118, 126-127, 129 151; *Codex Justinianus*/Code, 83-84, 88, 93-96, 126-127; *constitutio Omnem*, 93
Juvenal, 150-151, 153

Kalhu, (Nimrud in Iraq),16-17, 18
kalû (lamentation priest), 19, 23
Kandalanu, Babylonian king, 15
Khusrau I, shah of Persia, 163, 166
Khuzistan (Iran), 163, 173
Kûm Ishqâw, village, 101
Kunnash, 168

Lactantius, 78; *Divinae institutiones*, 78; *De opificio Dei*, 78
Law Code of Hammurapi, 17
Law of the Twelve Tables, 82
Leo, emperor, 123-127, 129-130; Law, 123-124, 126, 129
Lex Falcidia, 129
Livy, 67
Louis the Pious, king of France, 191-193
Lucan, 150-151, 153
Ludlul bel nemeqi, text, 17
Lugale, text, 17, 19
Luke, apostle, 25-26, 33, 155, 168
Lupus of Ferrières, 188, 192

Macrobius, 191
Manishtushu, Babylonian king, 17
Manuscripts, Besançon, Ms 594; Laon, BM, Ms. 444; Leiden,

UB, Vossius.lat.F. 48; Leiden, UB, BPL. 36, 87 and 88; Oxford, Bodl. Lib., Laud.lat. 118; Paris, BnF, lat. 8671; Vatican, BAV, Reg.Lat, 1987
Marcus Aurelius, 76
Marduk, temple, 11
Mari, city, 5
Mark, apostle, 25, 155, 168, 172-173
Martianus Capella, 66, 185-187, 190, 192; *De nuptiis Philologiae et Mercurii*, 185, 188, 190-192
Martin of Laon, bishop, 188, 190, 192
Masarjawaih, see Seharbokt bar Mesargis
mašmaššu, exorcists, 21
Matthew, apostle 25, 33, 155, 168
Maximilla, 47-48
Maximus of Tyre, 42-45, 48
Megas, 128
Menander Rhetor, 101, 104, 108, 112, 137, 148, 151, 153-154, 158
Menas, (half)brother of Dioscorus, 121
Menas, saint, 119
Mesargis, family, 168, 170
m[e]*shalim*, 33
Michael Psellus, 50
Mons Epiphanius near Diospolis, 148, 157
Montesquieu, 83, 90
Moses, 33
Mosul, town, 4
Mul.Apin, 6-7
Murbach, monastery, 181
Muridac, grammarian, 188
Musaeus, 103; *Hero and Leander*, 103
Mušezib, 23

Nabonasaros, king of Babylon, 5
Nabû ša Harê, temple, 14,
Nabû, god of art and writing, 14, 16
Nabû, temple at Kalhu, 16, 18
Nadere Reformatie, 135
Nagada, region, 155
Napoleon, 85
Naresh, city, 59
Narsai, school director, 165
Nebukadnezar, Babylonian king, 17
Nehar Pakod, city, 59
Nestor, 105-106
Nestorian physicians, 163, 168
Nicanor, Ptolemaic general, 143
Nidintu-Ani, son of Anu-uballit, 19, 23
Nineveh, 4-5
Nisibis, theological school of, 164-167, 169
Nonnus of Panopolis, 103, 105, 138, 147; *Paraphrase of the Gospel According to John*, 103, 138; *Dionysiaca*, 103, 138;
Novellae/Novels, 89, 94-96

Odyssey, 27
Olympiodorus, 114, 147
Omboi, city, 141
Origen, 155, 157, 177
Orion, grammarian, 161
Orosius, 77
Orpheus, 105-106
Oxyrhynchus, city, 146, 150, 152-157, 160

Palladius, 151, 158
Pamprepius of Panopolis, 103, 114; *Ekphrasis*, 103
Pandectae, 83
Panopolis, 148-150, 155-157, 160-161; monastery, 152
Paphienus, 111
Paphion; see Aphrodite
Paul, apostle, 33, 35
Peristera, sister of Dioscorus, 120, 116, 118
Peter, apostle, 33, 143-144
Petrarch, 79
Philemon, 151, 153, 156, 158
Philippus Arrhidaeus, Babylonian king, 21
Philo, 40, 42, 44, 45, 151; *De Somniis*, 42
Phoebammon, son-in-law of Apollos,

121
Phoebus, see Apollo
Pindar, 151, 153
Plato, 41, 43, 74, 151, 153; *Timaeus*, 43; *Phaedrus*, 44
Pliny the Elder, 63-64, 67, 69, 75, 77-78; *Naturalis Historia*, 63, 67, 75-78
Plotinus, 42
Plutarch, 42, 45, 50
Pomponius Mela, 67, 75-76
Pomponius, jurist, 81, 87
Porphyrius, 43; *Ad Gaurum*, 43
Post-Glossators, 98
Priscian, grammarian, 91
Prosper of Aquitaine, 177, 179; *Liber sententiarum ex operibus sancti Augustini delibatarum*, 179
Protoevangelium of James, 26
protokometai, 121-123, 128
Pseudo-Appolonius of Laodycia, 103, 105-106; *Paraphrase of the Psalms*, 103, 106
Pseudo-Dionysius of Halicarnassus, 104; *Tehknê peri tôn panêgurikon*, 104
Pseudo-Zacharias, 166; *Ecclesiastical History*, 166
ps-Menander gnomologies, 151
Ptolemy, 10
Pumbeditha, city, 59

Q Gospel, 25-26
Qashway, 166
quaestores, 118

Rabban Shabur, monastery, 173
Rav, 59
Remigius of Auxerre, 186, 190
Romanus Melodus, 151
Rufinus, 177

Šamaš-eṭir, son of Ina-qibit-Ani, 19, 23
Sanhedrin, 55-56
Sappho, 151, 153
Sargon II, Assyrian king, 16
Seharbokt bar Mesargis, 168-174; *Gannat Bussame*, 168, 171- 174
Seleucia-Ctesiphon, city, 164-165
Seleucus II, Babylonian king, 15
Seliba Zeka (Seharbokt), 168
Senuthes (Sinuthios), son of Apollos, 122-123
Sepphoris, city, 58
Sergius of Resh'ayna, translator, 164
Sergius, Nestorian metropolitan of Jundishapur, 164-165
Servius, 191
Shammai, 54
Shapur I, shah of Persia, 163
Shapur II, shah of Persia, 163
Šibqat-Anu, 19, 23
Sidonius Apollinaris, 111
Simplicius, 41; *In de Coel. comm.*, 41
Sinai, mount, 54, 57
Sin-leq e-unnini, *kalû* (priest), 19, 23
Sippar, city, 17
Solinus, 63-71, 73-79; *Polyhistor/ Collectanea rerum memorabilium*, 63-69, 70, 74-79
Solomon, king of Israel, 144
Sophocles, 151, 153
St. Catharine's Monastery at Sinai, 148
St. Riquier, monastery, 180
St.-Pierre in Corbie, monastery, 187
Strabo, 151, 153
Stratocles, 48
Suetonius, 177; *De viris illustribus*, 177

Tabernacle, 33
Tekrompia, cousin of Dioscorus, 118, 120-121
Temple of Jerusalem, 53, 55
Tertullian, 72, 74, 77; *Ad nationes*, 72; *Praescriptiones*, 72
Theocritus, 151, 153
Theodora, empress, 140
Theodore bar Koni, Nestorian exegetes, 169
Theodorus of Pentapolis, 110
Theodosius, emperor, 58, 84, 87;

Codex Theodosianus, 84
Theodulf, 182; *Libri Carolini*, 182
Theophilus of Antioch, 155
Thessalonians, 156
Thomas à Kempis, 73
Thucydides, 151, 153
Tiberias, city, 55, 58
Timothy I, Nestorian catholicos, 164-165
Titus, 156
Tribonian, officer, 88, 91
Tribunus, *archiatros*, 166
Triphiodorus, 103; *The Taking of Ilios*, 103
tupšar Enūma Anu Ellil, diviners, 10
ṭupšar Enuma Anu Enlil, ('astrologer'), 20, 23
ṭupšarru, scribes, 17
Tura, 148, 155, 157, 160
Twelve Tables, 82, 85

Ulpian, jurist, 81, 151
Uruk, city, 14-16, 19-23

Valentinian, emperor, 87
Venus, goddess, 5
Vergil, 150, 151, 153
Vetus Gallica, 181
Victor, priest, 123
Vincentius Lirinensis (de Lérins), 64-65, 70-79; *Tractatus pro catholicae fidei antiquitate et universitate adversus profanas omnium haereticorum novitates*, 65
Vision de Dorothée, 105
Voetius, Gysbertus, 135

Wadi Natrun, 155
Walter of Châtillon, 74
'War of the Ghosts', 28, 32
White Monastery at Panopolis, 148, 156
Wigbod, 181
Winibertus, student of John the Scot, 188

Yohanan, rabbi, 55

Zeus, 106-107
Ziggurat of Ur, 1
Zistaq, schoolmaster, 164